About the Authors

David Owen is the author of 19 fiction and nonfiction titles, including *Tasmanian Devil* and *Shark: In peril in the sea*. He is also the author of the popular 'Pufferfish' detective series set in Tasmania. He is the Official Secretary of the Governor of Tasmania.

David Pemberton is a wildlife biologist and former manager of the Tasmanian Government's Save the Tasmanian Devil Program. He has published scientific papers on a wide variety of conservation challenges, including seabird and seal bycatch in fisheries, and is co-editor of *Saving the Tasmanian Devil,* and co-author of *Tasmanian Devil* with David Owen.

'Educational and enlightening, and highly accessible to the reader' ***Aussie Reviews***

'Very well written and fascinating ... a compelling but tragic account of the complex cultural, political, economic, and ecological factors responsible for the extinction of the thylacine.' ***Conservation Biology***

'I would recommend it as an informative and readily accessible text well placed on any professional or home bookshelf.' ***Journal of Mammalian Evolution***

Note: *Tasmanian Tiger* was originally published under the title *Thylacine*. It was winner of the Royal Zoological Society's Whitley Award for Historical Zoology, 2004. This edition has been revised and updated.

Cover image: There is, of course, no extant colour photograph of a living Tasmanian tiger. The image on the c[illegible]s from black-and-white photographs [illegible]r pelts, using AI and Photoshop.

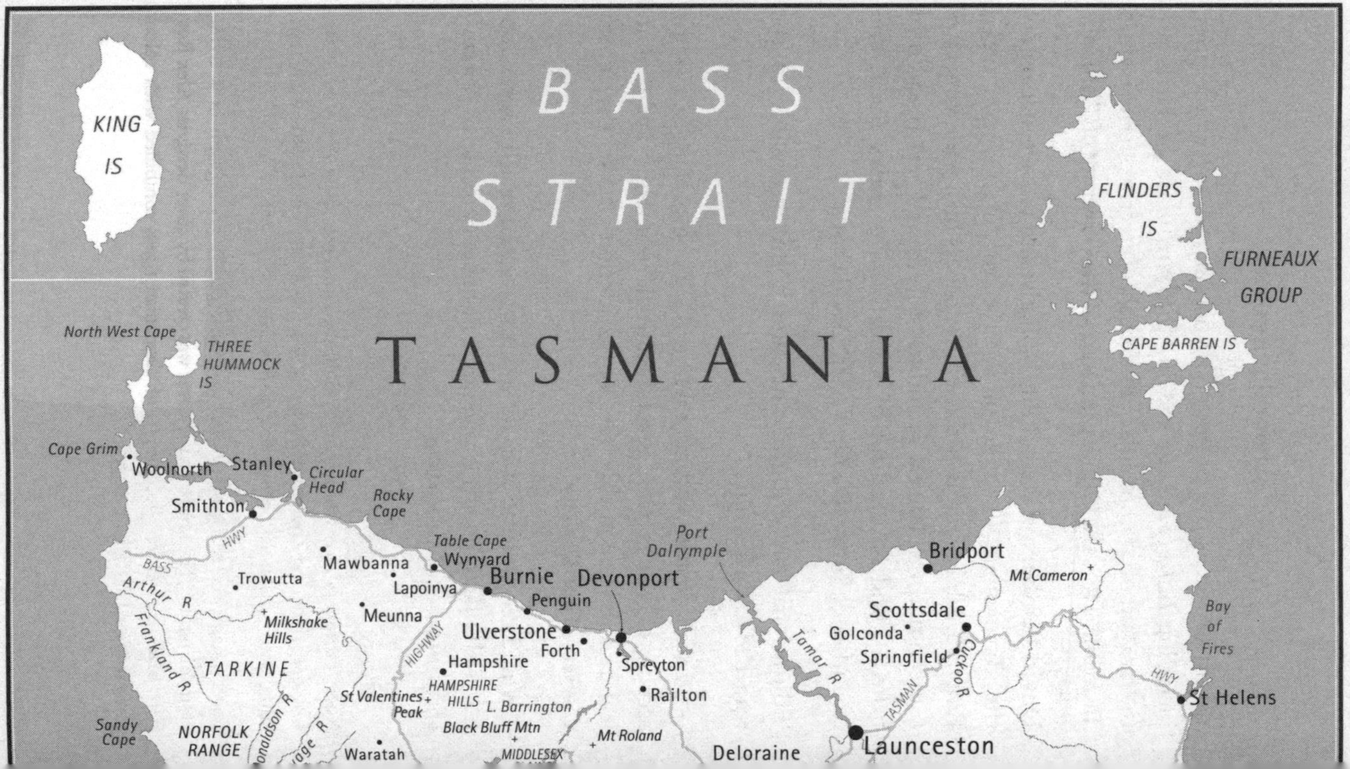

KING IS
BASS STRAIT
FLINDERS IS
FURNEAUX GROUP
CAPE BARREN IS
North West Cape
THREE HUMMOCK IS
TASMANIA
Cape Grim
Woolnorth
Stanley
Circular Head
Smithton
Rocky Cape
Table Cape
Wynyard
Port Dalrymple
Bridport
Mt Cameron
HWY
BASS
Mawbanna
Trowutta
Lapoinya
Burnie
Devonport
Penguin
Arthur R
Milkshake Hills
Meunna
Ulverstone
Forth
Scottsdale
Golconda
Springfield
Cuckoo R
Bay of Fires
Frankland R
TARKINE
HIGHWAY
Hampshire
Spreyton
Tamar R
HAMPSHIRE HILLS
St Valentines Peak
L. Barrington
Railton
TASMAN
St Helens
Sandy Cape
NORFOLK RANGE
Black Bluff Mtn
Mt Roland
Waratah
MIDDLESEX
Deloraine
Launceston

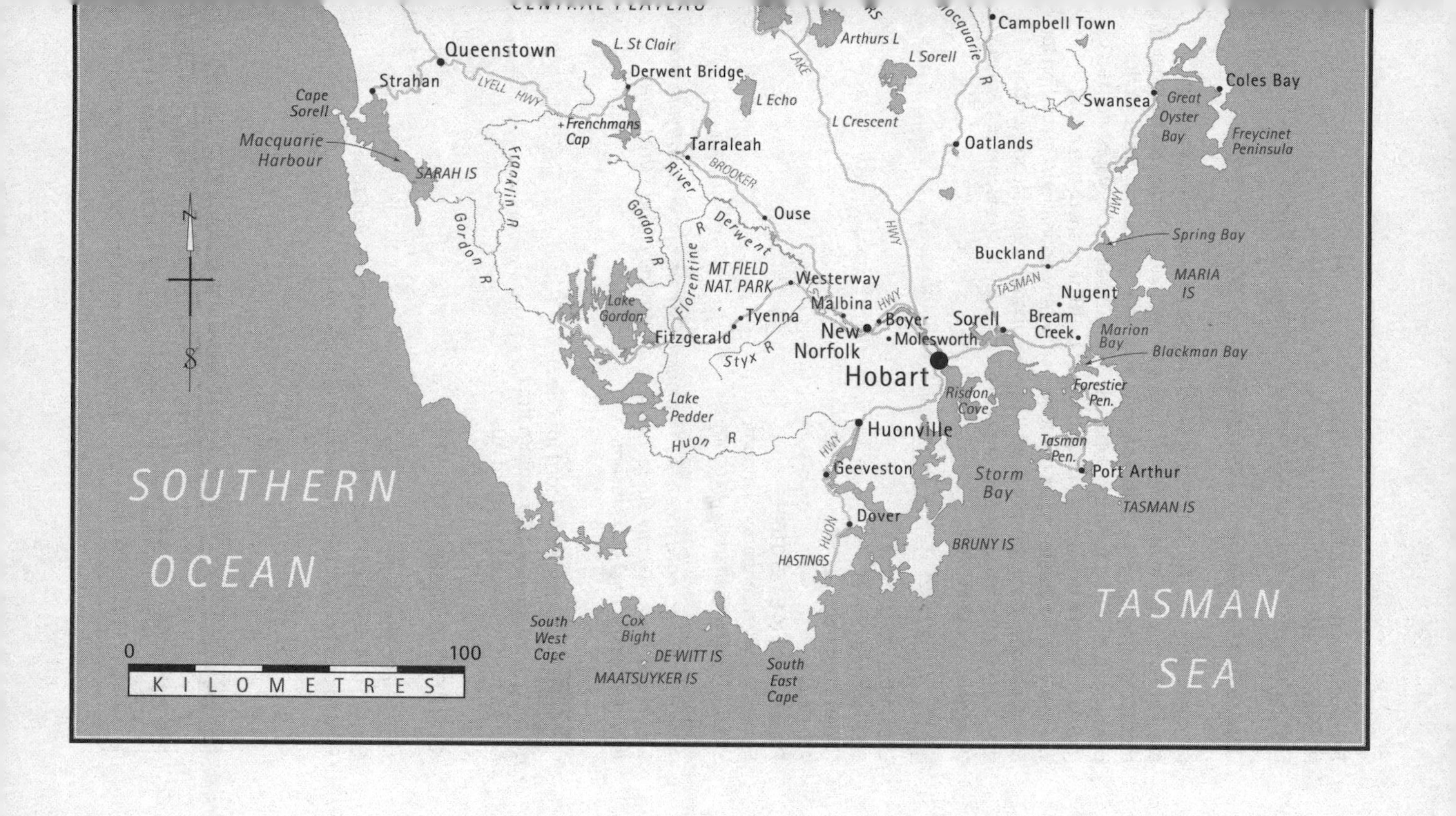

Campbell Town
Queenstown
L. St Clair
Arthurs L
L Sorell
Macquarie R
Strahan
LYELL HWY
Derwent Bridge
L Echo
Coles Bay
Swansea
Great Oyster Bay
Freycinet Peninsula
Cape Sorell
Frenchmans Cap
L Crescent
LAKE
Oatlands
Macquarie Harbour
Tarraleah
BROOKER
SARAH IS
Franklin R
River
Gordon R
Ouse
Derwent
R
HWY
Spring Bay
Buckland
MARIA IS
TASMAN
Florentine
MT FIELD NAT. PARK
Westerway
Nugent
Lake Gordon
Malbina
HWY
Tyenna
Boyer
Sorell
Bream Creek
Marion Bay
Fitzgerald
New Norfolk
Molesworth
Blackman Bay
Styx R
Hobart
Forestier Pen.
Risdon Cove
Lake Pedder
Huonville
Huon R
Tasman Pen.
Geeveston
Port Arthur
Storm Bay
SOUTHERN OCEAN
TASMAN IS
Dover
HUON
BRUNY IS
HASTINGS
TASMAN SEA
South West Cape
Cox Bight
DE WITT IS
MAATSUYKER IS
South East Cape
0
100
KILOMETRES

This second edition published in 2023
Published under the title *Thylacine* in 2011 in paperback
First published under the title *Thylacine* in 2003 in hardback

Allen & Unwin
Cammeraygal Country
83 Alexander Street
Crows Nest NSW 2065
Australia
Phone: (61 2) 8425 0100
Email: info@allenandunwin.com
Web: www.allenandunwin.com

Allen & Unwin acknowledges the Traditional Owners of the Country on which we live and work. We pay our respects to all Aboriginal and Torres Strait Islander Elders, past and present.

A catalogue record for this book is available from the National Library of Australia

ISBN 978 1 76147 039 4

Index by Puddingburn
Set in 11/14.5 pt Garamond 3 by Midland Typesetters, Australia
Printed and bound in Australia by the Opus Group

10 9 8 7 6 5 4 3 2 1

The paper in this book is FSC® certified. FSC® promotes environmentally responsible, socially beneficial and economically viable management of the world's forests.

DAVID OWEN AND
DAVID PEMBERTON

TASMANIAN TIGER

SECOND EDITION

The tragic story of the thylacine

ALLEN&UNWIN

For Leisha, Hilton and Larry

DO

For Rosemary, Sam, Elsa and Ollie
and all the grandchildren

DP

Contents

PREFACE TO THE SECOND EDITION

This book was first published in 2003 as *Thylacine: The Tragic Tale of the Tasmanian Tiger*, with David Owen as sole author. During the extensive research phase, wildlife biologist David Pemberton provided very significant advice and so it is appropriate and fitting that this revised edition is now co-authored.

The original publication won the 2004 Whitley Award of the Royal Zoological Society of New South Wales; one indicator that the lure of the mysterious thylacine endures, even given the fact that the island's unique top-order carnivore was callously hunted to extinction. The book's enduring popularity led to an invitation from Elizabeth Weiss, Allen & Unwin Publisher, to consider writing a revised edition. New material has been added; some existing text modified; and we are particularly pleased to include new, original reference to the art of the late Tasmanian Aboriginal artist Karen Casey, with thanks to her son Dan Young.

Is the Tasmanian tiger still 'out there'? In the 20 years that have passed since this book was first published, there have continued to be 'sightings'. But as natural history authors we consider that there is no chance the thylacine is not extinct.

The original Preface and Acknowledgements follow and are unchanged, being an integral part of this original work on Australian wildlife.

PREFACE TO THE FIRST EDITION

On 7 September 2002, Threatened Species Day was once again observed across Australia. In Hobart the blustery, windswept conditions didn't prevent stalls and exhibitions springing up on the lawns at Parliament House. Conservationists and politicians gave interviews for television and radio. The deliberately chosen date is a sorry one in Tasmanian history, for it was on 7 September 1936 that the last known thylacine, commonly known as the Tasmanian or 'Tassie' tiger, died in captivity at the Hobart Zoo, victim of the Depression, neglect and a century of deliberate species persecution.

In an unintended, eerie coincidence, the northern city of Launceston hosted an auction that very morning, the centrepiece of which was a rug made of eight thylacine skins. It had reputedly been purchased for three pounds early in the twentieth century, since when it had remained in private hands. Now the State's two main museums and a hotel group had jointly purchased the rug for over a quarter of a million dollars. Was it right to spend all that money on memorabilia of an extinct species on the very day devoted to raising funds to help endangered ones?

This is but one more controversial and sadly ironic chapter in the ever-expanding saga of the Tasmanian tiger, the mysterious marsupial predator that evolved over tens of millions of years and was hunted to extinction in the blink of an eye, because of its supposed attacks on sheep. But why is it that the longer the animal stays dead, the more we are fascinated with it?

Guilt, remorse and the tantalising possibility of its continued existence go some way towards providing an explanation. Arguments rage over its ability to have withstood its systematic persecution through trapping and snaring, poisoning and shooting. Even now, every third Tasmanian has a 'true' tiger-sighting story. Then there is the future. The Australian Museum's dedicated scientific team hopes to clone the thylacine by the year 2010. Preposterous or possible, no one yet knows. The only certainty is that the thylacine's story is not yet complete.

This book, one of just a handful devoted to the thylacine, is intended to be the most comprehensive yet in its coverage. Evolution and extinction, pre-European Tasmania (Trowenna), the rapid colonisation of the island with all its consequences, and the fierce conservation clashes of more recent times each have their place in the mystery-shrouded story. This nocturnal wolf/dog striped carnivore, with its huge jaws, continues to fascinate writers, artists, the true believers out there looking for it and those who, unexpectedly, become convinced they have seen one, crossing a rural road, disappearing into the bush at dusk, or trotting on a remote Tasmanian beach. All enrich the world of the thylacine.

The science of survival does not rate the animal's chances highly. But many people do, and Australia's large island state, with its great tracts of untouched wilderness and small human population, already has a strange and unlikely history, as these pages will show.

The thylacine *is* Tasmania. To that extent alone, it lives on.

ACKNOWLEDGEMENTS

This is a book of many voices, contemporary and long silent. Each one, no matter how modest, is part of the never-ending thylacine saga. My lengthy and varied research has particularly relied upon a number of individuals and institutions without whose generosity this book would not have taken the form presented here.

At some point in 2001 Richard Flanagan put me on the trail of the Tassie tiger—I am enormously indebted to him for this.

The staff of the Tasmanian Museum and Art Gallery (TMAG) have been unfailingly helpful over a period of two years. In particular, vertebrate curators David Pemberton and Kathryn Medlock, who oversee the world's most diverse collection of thylacine material—and who field innumerable thylacine requests from all over the world—have shared valuable information and provided invaluable critical analysis; I am grateful to them both, and haven't forgotten my beer promise. Particular thanks also to Jacqui Ward of the TMAG's Photographic Collection and to Director Bill Bleathman and Deputy Director Dr Andrew Rozefelds.

The thylacine existed safely in the island of Tasmania for tens of thousands of years prior to European settlement. A chapter in this book attempts to reconstruct something of that pre-1803 environment. I thank Greg Lehman, Assistant Director of the

University of Tasmania's Riawunna Centre for Aboriginal Education, for his generosity in this regard and Eve Mills, Senior Curriculum Officer, Aboriginal Education, Department of Education.

Nick Mooney, Nature Conservation Branch senior wildlife officer, has for many years been the official responsible for thylacine sightings and research. I am grateful to him for his ongoing sharing of his specialist knowledge.

Thanks also to Robert Paddle, both for the generous use I have been able to make of his landmark work on the thylacine and for his subsequent advice.

The research process was considerably aided by my having access to the Peter and Elizabeth Mercer Tasmaniana Collection, Jane Franklin Hall, Hobart.

Needless to say, in the end, where they may appear, any errors of judgement, argument or fact are exclusively mine.

I owe a significant debt to my publisher, Ian Bowring at Allen & Unwin, for commissioning this book; and it has been a great pleasure to work closely on the project with senior editor Emma Cotter.

I thank the following for granting interviews: Mike Archer, Garry Bailey, David Boon, Bob Brown, Geoff Law, Heather Rose and Steve Thomas.

The book is enriched by the addition of certain material, for which I thank Vita Brown, Jane Cooper (daughter of the late Jackson Cotton), Ian Faulkner, for his beautiful thylacine drawings, Carol Freeman, Pete Hay, Christine Lucas, Michael McWilliams, Daniel Moynihan, Jonathan Nadji, Ian Pearce and the staff of the Archives Office of Tasmania.

The first-person accounts of thylacines which head up each chapter are a significant feature of this book. These recollections were associated with the first book devoted exclusively to the thylacine: *Search for the Tasmanian Tiger* by Quentin Beresford and Garry Bailey, published by Dan Sprod's Blubber Head Press in 1981. (It was pleasing to be able to make contact with four of the 39 entrants, Geoff Aschman, Kath Doherty, Betty Holmes and Elizabeth Okines.) More details of the book are provided in Chapter 12.

Thanks also to Chris Tassell, Director, Queen Victoria Museum and Art Gallery, Malcolm Wells, Col Bailey, Jennifer Broomhead, Heather Felton, Georgiana Fien, Margaret Harman, Christine Holyoak, John Long, Richard Lord, Dianne Mapley, John Pemberton, Tracy Elizabeth Robinson, Michael Roe, Randy Rose, Emma Singer, Don Stephens, Ann Sylph, Christine Tarbett-Buckley, Margaret Valentine, Cate Weate, Gerard Willems and Richard Wilson.

Finally, much is owed to Eric Guiler for his lifetime's devotion to the Tasmanian tiger.

David Owen

1

WHAT'S IN A NAME?

> The rather timid-looking creature was biscuit coloured, and I immediately thought it was a Labrador dog. But there was something strange about the head and face which puzzled me. It was higher and wider across the forehead than a Labrador, and the face was longer and thinner. Perhaps it was just a 'bitzer', with some Labrador in it. I was quite alone as my companions had gone off to look for a track leading towards Macquarie Harbour. Not even a bird call, a creaking branch . . . disturbed the peace and quiet around us as we continued to stare at each other.
>
> ELIZABETH OKINES, SANDY BAY

Uncertainty, confusion and misinformation—deliberate or otherwise—have always been part of the baggage of discovery. Familiar names combined with the words 'false' and 'mistake' were often applied to the phenomena of the New World. NASA and other space agencies regularly lose, or fatally programme, exploration modules and equipment. Astronomers and astrophysicists are obliged to constantly contradict and overwrite existing theories. Palaeoanthropologists continue to backdate the origin of hominins. Through history human fallibility has, arguably, been the only constant.

Thus it was that in December 1642 Dutch mariner Abel Tasman's landfall on the south-east coast of the temperate island that now bears his name, although not actually a mistake, was supposed to have had an entirely different outcome. The economically greedy and commercially secretive Dutch East India Company, under its Batavia-based Governor-General Antony Van Diemen, fully expected Tasman to discover a vast land of great riches and fertility, inhabited by a civilised, friendly people. Its fabled existence had been the subject of European speculation for centuries. It was *Terra Australis Incognita*, the unknown south land, a land mass whose antipodean weight must balance the great Northern Hemisphere continents.

In the event Tasman and his two-ship expedition spent only a few days at anchor, from the first to the fourth of December, near Blackman Bay on Tasmania's Forestier Peninsula, before sailing further east to discover the islands that became New Zealand—which, he speculated, was the main continent of the unknown south land and might also be joined to Cape Horn.

On an interesting point of nomenclature, the vessels were the 60-ton warship *Heemskerck* and the smaller brig *Zeehaen*, which between them today have Tasmanian mountains, a town and a winery named after them. Tasman sailed aboard the *Heemskerck*, which was skippered by Ide Tjercxszoon. Had Tasman died, Tjercxszoon would have formally succeeded him as commander, which might have led to the island having an even stranger name.

The short stop on the Forestier Peninsula had four outcomes of note. First, a sailor swam ashore and planted a flag, thereby taking formal Dutch possession of the new land which they had already named Van Diemen's Land when at anchor off Macquarie Harbour on the west coast a week or so earlier.

Second, they had no conception that they had circumnavigated the southern half of an island, reckoning this coastline to be part of *Terra Australis Cognita*, the known south land, by then referred to as New Holland. The Dutch had already explored the coastline of New Holland from Cape York in the north down the western coast to the Great Australian Bight in the south. This misconception was to endure for a long time.

Third, a small exploration party searching for fresh water found little—despite the nearby presence of pleasant Bream Creek, which also happens to have a winery named after it—but they did hear human sounds, including music they described as like a trumpet or small gong, and they also found climbing notches cut into trees. They correctly surmised that the treetops contained food sources (mainly possums), but the wide spacing of the notches led them to wonder if perhaps giants inhabited the place. Abel Tasman's journal described the notches as 'about 5 foot asunder, so that we must either conclude that these people were very great, or else that they have some unknown trick'. Jonathan Swift's 1726 *Travels Into Several Remote Nations of the World by Lemuel Gulliver* is set in or near Van Diemen's Land and has, of course, the famous giant Brobdingnagians. Swift had access to an English translation of Tasman's journal. Is the greatest and most profound satire on the follies of humanity owed to a race that was destined for near-genocide?

And fourth, the party in search of water 'saw the footing of wild beasts having claws like a tyger, and other beasts'.[1] This is the first known written reference to a Tasmanian marsupial, and is generally accepted as recording the marks of a thylacine. Certainly, thylacines were plentiful in that area—the nearby steeply wooded Ragged Tier in particular, as well as the Spring Bay area, later to be intimately associated with a government bounty to exterminate the animal—but those prints near the turquoise waters of Marion Bay could just as well have been from a wombat, which has a large pad and elongated, distinctive

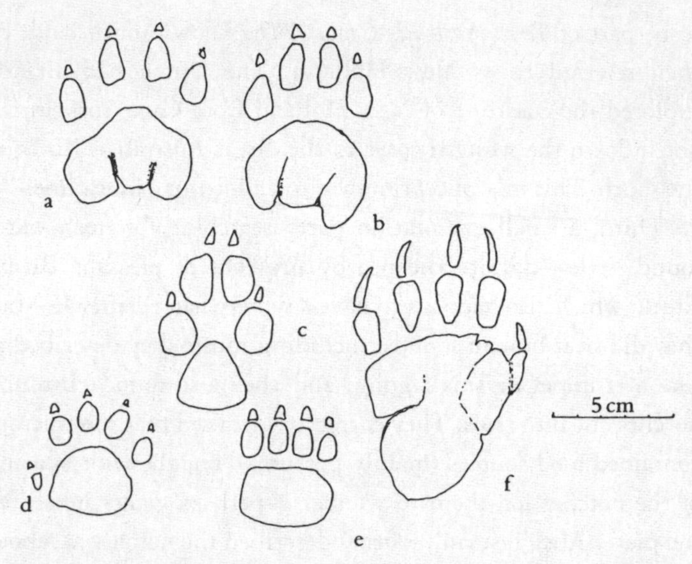

Footprints of thylacine, dog, devil and wombat. ***a*** *left front foot of thylacine;* ***b*** *left rear foot of thylacine;* ***c*** *dog;* ***d*** *right front foot of devil;* ***e*** *rear foot of devil;* ***f*** *left front foot of wombat.*

claws—very tiger-like. The fact is that once Europeans sighted the thylacine, with its remarkable stripes, the erroneous name originally applied by the Dutch explorers stuck.

Uncertainty, confusion, misinformation: when a rare species becomes victim of all three, over a sustained period of time, its chances of indefinite survival are slim indeed. The thylacine's uniqueness in part proved its undoing, because the earliest Van Diemen's Land European settlers, 150 years after Tasman, had to 'invent' it—and they did so in a welter of confusion, understandably if wrongly ascribing to it the characteristics of known predatory mammals. In this way it became a big cat/wolf/wild dog/ hyaena hybrid, an elusive New World creature as disturbing as the venom-spurred platypus was bizarre.

Immediately prior to the British establishing a precarious settlement at Risdon Cove in the island's far south in late

What is it? Early Van Diemen's Land settlers and scientists were perplexed by the weird marsupial carnivore, with its apparent cat-like, dog-like, wolf-like and hyena-like characteristics. (Robert Paddle, *The Last Tasmanian Tiger: The History and Extinction of the Thylacine*, p. 53)

1803, a French expedition led by Captain Nicolas Baudin had extensively charted parts of the island's coastline. Although political motives were ascribed to this voyage—France and England were at war—its *raison d'être* was scientific. Many thousands of biological specimens were recorded and interaction with the indigenous peoples was frequent and, mostly, amicable. From them the French learnt a great deal. Yet there is not one reference to thylacines. While this may seem an unaccountably odd omission, given the great variety of animals, birds, fishes and insects described in Baudin's journal, it is, perhaps, an early pointer to this predator's elusiveness and nocturnal habits—two known characteristics which today lead some Tasmanians to the conviction that the thylacine yet exists.

Geopolitical ambitions of the day notwithstanding, zoology, botany, mineralogy and anthropology had become of central importance to European voyages of exploration and settlement. A massive increase in knowledge was radically changing scientific disciplines in Britain and continental Europe, which had for so long been fixed in dogma. Thus, as an example, Aristotle's faunal catalogue *Historia Animalium* was in some senses only superseded by the monumental 1735 *Systema Naturae* of the Swede Carolus Linnaeus.

In 1806 the island's Deputy Surveyor-General, George Harris, was the first to scientifically classify the 'tyger'. He named it *Didelphis cynocephala*, the first Latin term placing it in the sub-order of South American marsupials, the second meaning 'dog-headed'. But this was found to be incorrect by, horror of horrors, French scientists following the groundbreaking work of their compatriots Georges Cuvier and Geoffroy Saint-Hilaire, who had earlier redefined zoology by introducing the science of comparative anatomy. This meant that classifications were to be established, and when necessary re-established, on the basis of anatomical relationships. In this way Cuvier and Saint-Hilaire set up the fundamental genera and classification schemes still in use. The animal so confused science that its classification changed three times until the current *Thylacinus cynocephalus* was arrived at in 1824.[2]

So much for science. But what were the early perceptions of the convicts and the settlers slowly spreading out from Hobart Town and Port Dalrymple (Launceston), who almost without exception had no knowledge of or interest in the nature of this creature?

Interactions with it were infrequent. Yet it soon developed a fearsome reputation, well-founded or not. In 1822 the Surveyor General, George William Evans, published an account of the island, in the zoology section of which he refers to the 'opossum-hyena' that 'few . . . have seen'. In somewhat contradictory fashion, Evans further observed that:

> this animal of the panther tribe . . . though not found in such numbers as the native dog is in New Holland, commits dreadful havoc among the flocks. It is true that its ravages are not so frequent; but, when they happen, they are more extensive. This animal is of a considerable size, and has been known, in some few instances, to measure six feet and a half from the tip of the nose to the extremity of the tail. Still it is cowardly, and by no means formidable to man: indeed, unless when taken by surprize, it invariably flees from his approach.[3]

So even after twenty years, the creature remained ill-defined in the popular imagination, as well as the scientific mind. This is amply demonstrated by the multitude of names it acquired.

The names known to have been bestowed on the thylacine include: zebra opossum, zebra wolf, Tasmanian zebra, marsupial wolf, striped wolf, tiger wolf, Tasmanian wolf, Van Diemen's Land tiger, Tasmanian tiger, bulldog tiger, greyhound tiger, hyaena tiger, dog-faced dasyurus, dog-faced opossum, hyaena, native hyaena, opossum-hyaena, dingo/Tasmanian dingo and panther. Aboriginal names will be addressed in a later chapter.

This list may seem improbable and suspiciously long, but the explanation is relatively simple. Genuine confusion and uncertainty aside, the sad fact is that the thylacine attracted names of opprobrium from human forces increasingly opposed to its existence. (Unlike, for example, the herbivorous and generally docile giraffe, which the ancient Romans called *camelopardalis*: the camel marked like a leopard.)

University academic Robert Paddle, whose landmark work *The Last Tasmanian Tiger* is scholarly, original and insightful, makes the point that the common, but incorrect, translation of its classification name *Thylacinus cynocephalus* as 'the dog-headed pouched-dog' is not just inelegant, but 'borders upon the stupid and crass'.[4]

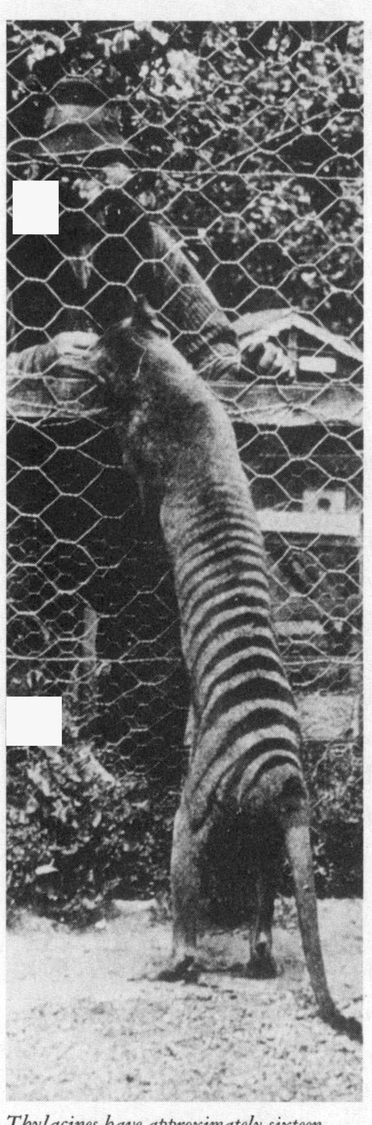

Thylacines have approximately sixteen stripes. This is one of very few photographs showing thylacine–human interaction. Is it being fed? Petted? (Collection Tasmanian Museum and Art Gallery)

Other translations, such as Australian zoologist and mammologist Ellis Troughton's 1941 version, seem little better: 'the pouched dog with the wolf head'. Troughton, for all his experience, prolonged an unsubstantiated belief. He wrote: 'The thylacine is usually spoken of by country folk as the "Tiger" in reference to the sixteen to eighteen dark chocolate stripes across the lower back, which provide a similar protective camouflage to that of the tiger when moving amongst foliage barred by sunlight'.[5] A presumptive statement of fact—and who, perhaps, can blame the author?—it nevertheless was unhelpful, continuing to rely on a cross-species assumption. Tiger stripes, which are broken striations, and thylacine stripes, which are firm stripes, don't necessarily have the same function.

What did the word 'tiger' mean in the consciousness of the early nineteenth-century Britons and Irish transplanted to remote Van Diemen's Land? Colonial experience of, for example, Bengal tigers, meant stealth, ferocity, cunning, near-invisibility, powerful swimming, nocturnalism, awesome strength and, from time to time, human-eating. What fledgling Vandemonian sheep farmer would happily go to bed with that on his mind? An accurate examination of the thylacine's predatory habits shows that besides nocturnalism it has virtually none of those tiger-like characteristics. (Renowned Indian tiger hunter, naturalist and author Jim Corbett in the 1940s wrote that 'a tiger is a large-hearted gentleman with boundless courage and when he is exterminated—as exterminated he will be unless public opinion rallies to his support—India will be the poorer by having lost the finest of her fauna.'[6])

The terms 'bulldog tiger' and 'greyhound tiger' were scientifically applied in a short-lived attempt to prove that the island was inhabited by two species of thylacine, an elementary error caused by confusion over relative size differences between males and females, adults and juveniles. In the early 1920s anti-thylacine forces tried to resurrect this false notion in order to suggest thylacine plenitude.

The simplistic logic of applying the tag 'zebra' to the animal is also understandable. Those earliest settlers and their convicts would not have had any concerns about accurate naming of flora or fauna, not when starvation and unpredictable weather combined to obstruct the development of muddy Hobart Town. But the sheer inappropriateness of the connection is worthy of comment. Does the zebra, like the tigers of Asia, have stripes to blend in? Not with the African landscape! There has long been genuine uncertainty about this most distinctive of mammalian patterning. Perhaps the zebra has stripes to blend out, thereby aiding predators to identify the weaker individuals in the herd and saving all concerned unnecessary energy expenditure, for as soon as a kill is made the herd stops running.

Behaviourist Desmond Morris, a former Curator of Mammals at London Zoo, identified nine possible theories to explain the stripes of the zebra, ranging from safety in mass optical illusion through personal identification to an ingenious cooling system. The original camouflage explanation is discredited: 'Their natural history does not fit with the explanation that has been repeated in textbooks for decade after decade.'[7] Research also suggests a 'barber-pole' effect of thin black stripes effectively acting as a deterrent to flies and other pests: 'Horseflies are averse to landing on striped objects . . . This suggests that any hoofed animal that reduces its overall dark outline against the sky will benefit in terms of reduced ectoparasite attack.'[8]

So, if the widely studied zebra has been a longtime victim of misinformation, what chance of a fair hearing for the elusive Tasmanian thylacine, the 'zebra wolf'? Again, there is a certain logic in the wolf attribute, given the doglike features of the thylacine, its carnivorous nature and other features pointing to convergent evolution in the two species. But although there is no link to the canid family, the thylacine suffered through its perceived association with *Canis lupus*, the grey wolf of the Northern Hemisphere. (One striking difference is that dogs have six incisors; thylacines have eight.)

Few other animals have been so unfairly reviled and wrongly feared as the wolf. In the Middle Ages people identified as werewolves were burnt at the stake, having first been gruesomely tortured. Infection with rabies through the bite of a rabid domestic dog may have led to such identification, since both animal and human victims of the deadly virus salivate extensively and can exhibit uncontrollable, vicious behaviour as the brain becomes affected. The persistence of the European myth that wolves ate children found an echo in Van Diemen's Land. And, of course, the grey wolf is the evil protagonist of two particularly famous fairy tales.

Intensive research confirms that the belief in deliberate wolf predation upon humans is incorrect. According to a study reported by the International Wolf Center:

> In Europe and North America we only found evidence for 12 [non-rabid] attacks (with 14 victims) of which two (both in North America) were fatal, across a period of 18 years. Considering that there are close to 60,000 wolves in North America and 15,000 in Europe, all sharing space with hundreds of millions of people, it is apparent that the risks associated with a wolf attack are above zero, but far too low to calculate.[9]

As if all this was not enough, the wolf was and is blamed for livestock kills in those parts of Italy, Spain, Portugal, Greece and Mexico where it still exists, even though feral dog packs are the main culprits. Wolves are thought to prefer their natural prey, but the decimation of smaller mammal populations has led to an increasing dependence on livestock. Likewise, it is certain that the thylacine as a specialist predator would have an overwhelmingly natural preference for native prey, specifically kangaroo and wallaby.

Similar scapegoating occurs with canids elsewhere. Thus, southern African farmers of the valuable caracul sheep sometimes enclose pastures to prevent attacks by the black-backed jackal. 'Even so, the extent of predation by jackals here is probably exaggerated, with many of the sheep being killed by domestic dogs.'[10]

It seems inevitable that the island's elusive marsupial predator would eventually be demonised as a sheep taker. Destructive nineteenth-century bounty schemes, set up to save sheep from the thylacine, effectively led to its extinction. Today Tasmania's sheep farmers still suffer stock losses to other animals—but no-one's blamed the thylacine for a long time.

The thylacine's already terrible image was irredeemably tarnished once it attracted the hyaena tag. Although the two have no current scientific relationship, they share an intriguing biology: just as the world's largest contemporary marsupial carnivore defies neat categorisation, so too is the hyaena a mystery, being part-dog, part-cat and part-mongoose. Like the thylacine, it has its own classification (family Hyaenidae)—and: 'If any animal was less popular than the hunting dog in Africa during the days of the great white hunters, it was the hyena. Universally regarded as cowardly, loathsome, filthy and a dozen other undesirable adjectives, it too was persecuted as vermin . . . Along with vultures and dung beetles, it was regarded as a necessary evil'.[11]

The hyaena and the thylacine have very little in common, despite the superficial similarity in the sloping rear quarters. The latter is permanent in the hyaena, which has shorter back legs than forelegs, but the thylacine's marsupial ability to flatten its ankle joint into an elongated foot can achieve the same lowering effect. Furthermore, thylacines were reported to have 'a slow skulking habit. This suggestion is supported by those persons fortunate enough to have seen thylacines hunting'.[12]

Needless to say, there are plenty of differences between the thylacine and the hyaena, one being the former's modest vocalisation, so different from the scary 'mad laughter' of the hyaena, which in African folklore is associated with the witches of the night. At least the thylacine has been spared that. Anatomically it has been shown to most closely resemble a leopard, in that they both have short legs and long tails. But the hyaena link persists.[13]

Calling a thylacine a dingo isn't illogical either: both are champion carnivores uniquely identified with the vast Australian continent, although the ability of the latter to have some affinity with human beings perhaps spelled the end of the former. It is instructive to bear in mind that dingoes have long been regarded as a terrible pest to Australian farmers, because of their predatory habits and canine ability to take stock with relative ease.

The marsupial thylacine is the odd one out in that predatory threesome. Yet in the 1880s, when the Tasmanian parliament was in ferment over whether or not to introduce a bounty scheme to eradicate the thylacine, it was named by the pro-kill lobby as the Native Dingo. Again—if it hadn't been assured many times already—the thylacine's fate was sealed. It was almost everything in mammal form that people despised, feared, hated.

Now its loss is mourned. But that is not to say that there are not remnant believers in its continued cryptic existence.

2

IN THE BEGINNING: EVOLUTION

> One of the favourite stories told by bushmen is that if you catch a tiger by the tail he cannot touch you because he has a stiff back, but as far as I am concerned I would prefer that someone else held the tail. He is a very prettily striped animal and I feel it is a great pity they are all gone.
>
> ARNOLD PENNEY, DEVONPORT

The recent and brutally swift demise of the thylacine, a human-engineered folly, is brought even more sharply into focus when considering the length of time it took to evolve as Australia's largest extant marsupial carnivore. Its origins are shrouded in mystery, partly through lack of study.

The Australian continent began to part from the super-continent Gondwana as much as 200 million years ago, becoming fully separated about 50 million years ago and taking with it primitive marsupial stock, remnants of the earliest mammals that first appeared as small nocturnal creatures during the dino-saur-dominated Triassic, Jurassic and Cretaceous periods. A look at the dynamics of evolution will show how the thylacine eventually came to occupy its unique and seemingly too vulnerable niche.

It's generally accepted that Earth is about 4.6 billion years old, having been born out of a hydrogen-like solar nebula as a mass that condensed to a protoplanet happening to contain carbon-based chemicals, in time generating primordial life. The earliest evidence of this is found in 3.5 billion-year-old stromatolite rocks in Western Australia's Pilbara region. (A 2016 study controversially backdated earliest life to 3.7 billion years in Greenland rocks.) As described by University of New South Wales geologists Raphael Baumgartner and Martin Kranendonk, working with WA colleagues, strongly sulfidized stromatolites 'show textural features that are indicative of biogenic origins . . . dominated by petrographically earliest, nano-porous pyrite . . . consistent with a formation via sulfidization of [a] relationship with microbial communities.'[1]

In time, the single-cell organisms that first sprang up in Australia billions of years ago developed a nucleus, the protective repository of deoxyribonucleic acid, aka DNA, the building block of replication which is so exciting cloning advocates early in the twenty-first century. Indeed, thylacine DNA is being used in a major and controversial Australian thylacine cloning project (discussed fully in Chapter 14).

Elementary sexual reproduction was the next step along the evolutionary path. Many life forms so created were still incredibly small. Yet it took the great bulk of the Earth's known age for life to advance to that stage. This is known as the Cryptozoic.

The Phanerozoic is charted to include the present and can be said to have commenced about 570 million years ago. It is defined on the official Geologic Time Scale as being the time of 'evident life', thus distinguishing it from the Cryptozoic's 'hidden life'.

Earth's timelines are as scientifically intricate but as continually shifting as are the classifications of living things. One is

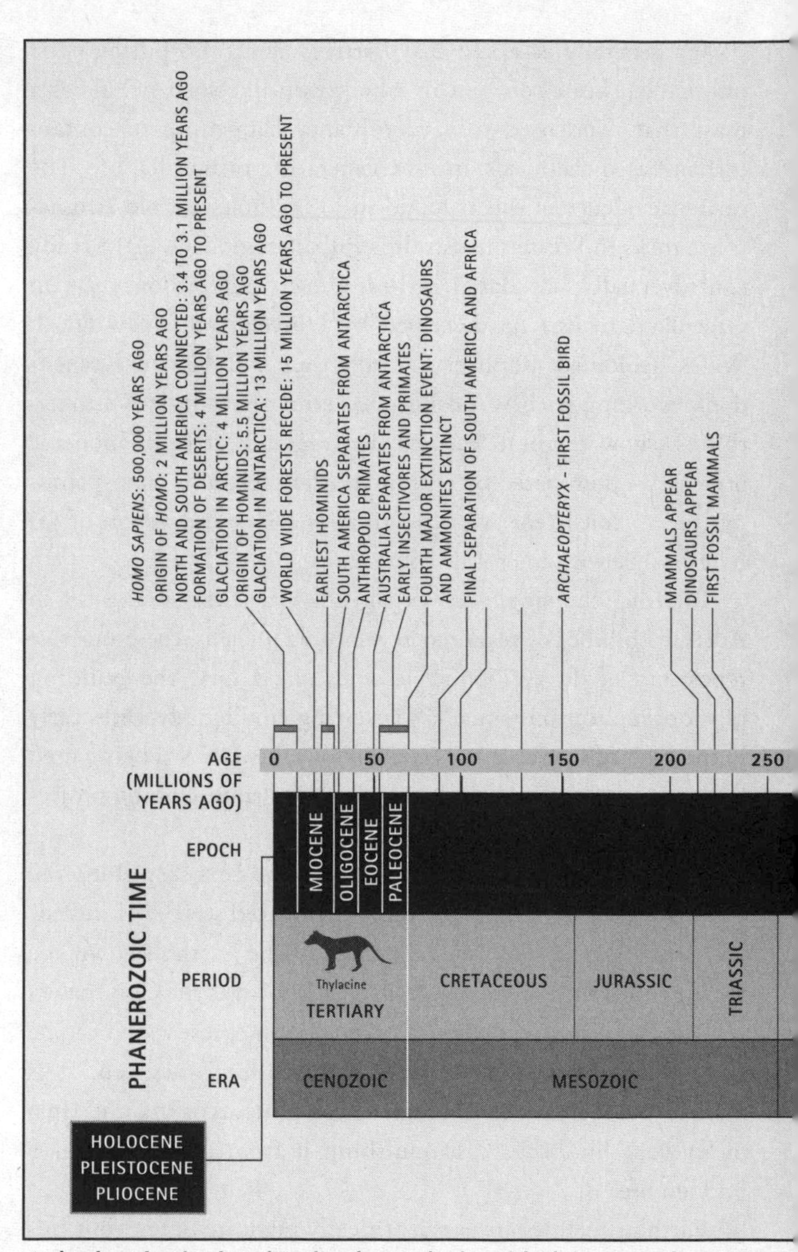

Archaeological and geological timeline showing the place of the thylacine in prehistory.

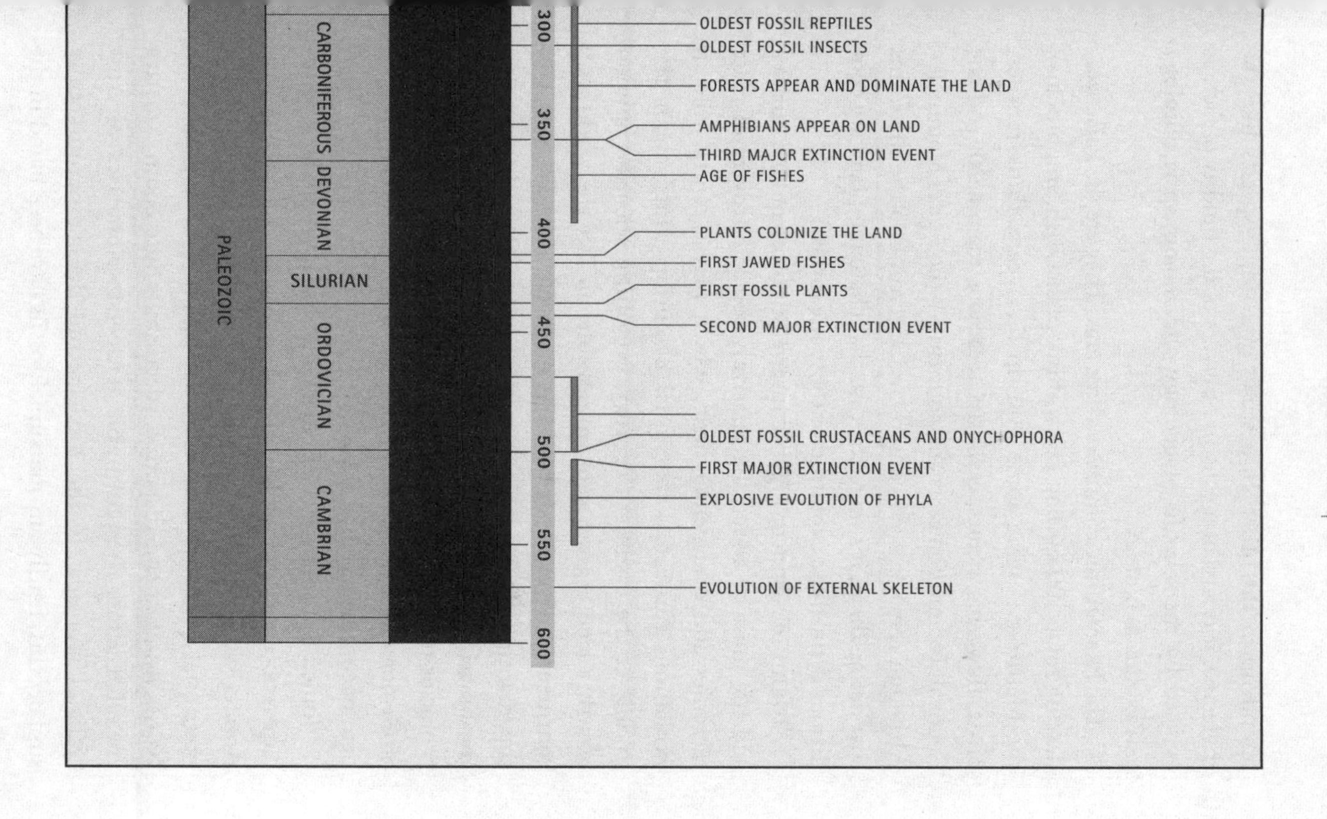
PALEOZOIC
CARBONIFEROUS
DEVONIAN
SILURIAN
ORDOVICIAN
CAMBRIAN
300
350
400
450
500
550
600
OLDEST FOSSIL REPTILES
OLDEST FOSSIL INSECTS
FORESTS APPEAR AND DOMINATE THE LAND
AMPHIBIANS APPEAR ON LAND
THIRD MAJOR EXTINCTION EVENT
AGE OF FISHES
PLANTS COLONIZE THE LAND
FIRST JAWED FISHES
FIRST FOSSIL PLANTS
SECOND MAJOR EXTINCTION EVENT
OLDEST FOSSIL CRUSTACEANS AND ONYCHOPHORA
FIRST MAJOR EXTINCTION EVENT
EXPLOSIVE EVOLUTION OF PHYLA
EVOLUTION OF EXTERNAL SKELETON

bound up in the other. In a study of a single species like the thylacine, it is advantageous to be aware of the fluidity of fact, as evinced by the scholars of geology and zoology, palaeontology and evolution biology.

The Phanerozoic comprises three eras. They are the Palaeozoic (lasting some 300 million years: think plants, fish, amphibians); the Mesozoic (lasting some 200 million years: think reptiles, birds, flowering plants, earliest Cretaceous mammals); and the Cenozoic (65 million years and counting—the Age of Mammals). Although these eras are constructed according to the life forms that define them, it's more the case that changing planetary conditions defined the evolution of those life forms.

To this self-evident organic interrelationship of land, air, sea and life must be added the relatively recent discovery of plate tectonics and continental drift, which play a central role in explaining Australian marsupial evolution. Until about the middle of the twentieth century, prevailing scientific opinion was that land masses were immovable. In fact the world's six continents have variously been one megacontinent, Pangaea, two supercontinents, Gondwana and Laurasia, and numerous other combinations, all of which have played definitive parts in the evolutionary jigsaw. (Samuel Warren Carey, founding professor of Geology at the University of Tasmania, was an early proponent of continental drift theory.)

Joined to comprise a shifting Gondwana were today's South America (once land-bridge linked to North America), Africa, Antarctica, India and Australia. This is thought to explain the marsupial journey—that the marsupials that originated in North America and dispersed successfully into South America travelled across then-habitable temperate Antarctica to Australia. Unfortunately the fossil record doesn't unequivocally support this; hardly any marsupial fossils have been found in the now frozen continent, which nonetheless as part of Gondwana hosted a great variety of vertebrate life.

Is the thylacine genetically linked to those earliest North and South American marsupials? Certainly, carnivorous predators evolved in South America, once that continent itself became isolated about 40 million years ago. Its family Borhyaenidae ranged in size from mink to bear, animals with large teeth and short limbs. Deriving from those specimens were the sabre-toothed marsupials: large, speedy and with powerful sets of teeth. Interestingly, the long-extinct Australian marsupial lion *Thylacoleo carnifex* (some believe it still exists!), apparently descended from either pygmy possum or wombat stock, had equally fearsome teeth: very large incisors and blade-like cutting carnassials. These teeth 'represent the most extreme specialisation of any known mammalian carnivore'.[2]

Prior to the discovery of continental drift, Australia's marsupial presence was explained with a theory of island-hopping from Asia. Then in the 1920s it was proposed that Antarctica might have acted as a land bridge. In 1981 a 40 million year old marsupial fossil was found on the Antarctic's Seymour Island. Belonging to a marsupial family previously known only in South America, it was a hugely significant find, all but proving the land bridge theory. (Subsequent fossil mammal fragment discoveries backdate the land bridge to over 55 million years.)

Recent discoveries confirming the existence of marsupial fossils in the Gobi Desert have complicated matters. The fragments were first discovered at Mongolia's Ukhaa Tolgod fossil site in the 1920s by American explorer Roy Chapman (upon whom Indiana Jones is modelled), but it was not until the 1990s that the tiny dinosaur-era mammal *Deltatheridium* was positively identified as a very remote relative of kangaroos and possums. Does this suggest revisiting a possible Asian origin for the marsupial?

There are also theoretical proponents for marsupials originating in Australia, not colonising it. Why not? Might there even have been primitive placentals on the Australian continent, banished by their more successful marsupial relatives?

Whatever the direction of migration, tectonic plate activity governed much of evolution. Features of continental movement include orogenesis (mountain building), the creation and extinction of oceans and seas, and great subterranean upheavals causing mass volcanic activity and resultant atmospheric change with huge increases in carbon dioxide levels. Such an increase in carbon dioxide occurred during most of the Mesozoic era, when Pangaea was breaking apart. It created a greenhouse effect which was ideal for large cold-blooded reptiles.

During the earlier Palaeozoic era, plants had become the earliest land colonists. They pumped oxygen into the atmosphere, thus offsetting the danger of ultra-violet radiation and enabling some of the fish to evolve safely into amphibians. These in turn became the earliest reptiles, which split into two lines, the petrolacosaurs (becoming dinosaurs and birds) and the pelycosaurs (leading to protomammals) of the warm Carboniferous and Permian periods.[3]

Dinosaurs were the dominant life form for 200 million years. Their sudden demise led directly to mammal domination. What happened? The catastrophic extraterrestrial theory involving asteroid or comet impact seems to provide the most likely answer. An indentation clearly visible near Mexico's Yucatan Peninsula, in the Caribbean Sea, as well as the presence of much extraterrestrial iridium dating to this point in rock strata around the world, is evidence for the theory. Such an impact would have caused great floods and firestorms and blocked out sunlight to large parts of the Earth for months. Loss of warmth, and the sunlight required for plant photosynthesis, suggests considerable immediate devastation. Unlike the large dinosaurs, the small, mostly nocturnal, heat-generating mammals then present were less vulnerable.[4]

Another theory holds that plate tectonic activity put so much volcanic particle debris into the atmosphere that a nuclear winter took hold. Sunlight would have been blocked and the food chain disrupted, terminating plant life and therefore the lives of the herbivorous dinosaurs upon which the carnivorous ones preyed. Whatever the truth, this momentous evolutionary event at the end of the Mesozoic left a world for the taking.

With the reptile dinosaur line now reduced to a pool of lizards, snakes, tortoises and crocodiles, mammals were in a prime position to diversify. By intriguing coincidence, this was round about the time that Australia set off on its long journey of isolation with its cargo of early marsupials and established monotremes. Isolation may well be a key term in explaining how the thylacine came to occupy its niche at the top of the food chain, as one of only a few carnivores.

By comparison, the South American experience was one of isolation interrupted. It seems that all of its mammals—the small herbivorous marsupials, the larger but slower carnivorous marsupials Borhyaenidae (squat, heavily built predators) and Thylacosimidae (sabre-tooth cats), and the placental ungulates (hoofed herbivores)—fell victim to what is known as the Great Interchange. This occurred when the sea level dropped and the Panama isthmus once again created a land bridge with North America. Faster, more efficient placental predators flooded in and decimated the South American endemics. That, of course, did not happen to Australia's evolving marsupials.

Reference has been made to confusion over the exact lineage of the thylacine. Carnivorous groups moving through the supercontinents left empty of dinosaurs included cat-like oxyaenids, dog-like mesonychids and mongoose-like hyaenodonts, either or all of which might lay claim to being remote thylacine progenitors.

Unfortunately, the Australian fossil record doesn't extend back beyond about the halfway point in the continent's isolation. This means that the thylacine's true origins may always remain unknown; although Mike Archer's palaeontological research has determined that the thylacine may have shared an ancestor with the borhyaenids, which had thylacine-like teeth and some skull characteristics of didelphid marsupials.[5]

Australia's combination of flat topography and prolonged subterranean chemical activity have worked against the preservation of fossils. Before the 25 million year mark, the continent's terrestrial vertebrate fossil record is meagre. Many assumptions have had to be made. But there is a spectacular exception with the presence of the monotremes, the egg-laying, milk-suckling mammals, represented today by the terrestrial echidna and the aquatic platypus. Their singular ancestral line bypasses all other mammalian branches as it leads directly back to the protomammals. To call monotremes 'primitive' may be taxonomically correct but, like the crocodilians, they have survived and flourished remarkably, surely a true test of Darwinian fitness.

As noted, marsupial mammals generally have been considered 'less advanced' than placentals. The logic supporting this—that placental gestation operates at a higher evolutionary level—ignores the circumstances of marsupial evolution. Once cut adrift, marsupial herbivores, carnivores and omnivores adapted in two ways, to each other and to the constantly changing island continent, from the extremes of its Antarctic position, which included months of perpetual southern darkness, to the periods of great inland seas, when the continent was like a patchwork of islands, to the relatively recent aridification of the centre contrasting with the humid forests fringing the north and north-east.

True marsupial diversification and radiation[6] is first evident in the Oligocene. But because fossil evidence of *Thylacinus cynocephalus* doesn't predate this, there is at present no way of determining its Australian lineage—let alone a link to the South American Borhyaenidae, to which it bears a superficial resemblance. What is clear is that the thylacine became widespread across the continent, which includes New Guinea. Its predecessor by some 4 to 6 million years was the considerably larger *Thylacinus potens*, whose fossil record is very meagre. The modern thylacine outlived both the marsupial lion, *Thylacoleo carnifex*, and the giant predatory kangaroo, *Propleopus.* It would also have competed to some extent with reptilian carnivores like the giant goanna and the terrestrial crocodile *Quinkana*. Even so, these represent a paucity of major predators, given the abundance and variety of herbivores. Down in size and occupying different niches are the Tasmanian devil, also once widespread on the mainland (and much larger than its modern-day descendant), and the tiger cats.

The Riversleigh fossil collections of Queensland provide a comprehensive thylacine picture back to a point: 'Five species ranging in size from that of a big domestic cat to that of a dingo thrived 25 million years ago—a sure indication that they had been evolving for a long time. But by 15 million years ago they were reduced to two species, and by 8 million years ago there was only one.'[7]

It is possible that the niche-specific thylacine developed relatively free of competition even from the much larger carnivorous megafauna, although rather than being an evolutionary bonus this may have spelled long-term problems once dingoes, men and dogs came on the scene. Although the theory of convergent evolution suggests that its canid-like features and some behavioural characteristics gave it a dominant position equivalent to that of the Northern Hemisphere wolf, some critical features may have been absent.[8] The thylacine may not have needed to develop genuine speed to go with its stamina when hunting wallaby and kangaroo. It may also have enjoyed the luxury of being able to depend exclusively upon these and similar prey, to the extent of feeding selectively from fresh kill. Perhaps it lacked the ability to scavenge and to learn to adopt an omnivorous diet in the face of later competition.

That the thylacine inhabited much of the Australian continent is further evidence of a lack of major competing predators; it also reveals an ability to adapt to dissimilar environments. This was particularly the case from about 6000 years ago when the interior began to dry out. There is no real evidence that thylacines were affected by the increasing aridity. On the contrary, Aboriginal rock and sacred cave art dating back to that time, in such areas as the Kimberley and Arnhem Land, vividly portrays the animal. There is also a famous mummified thylacine (see plate section), found in 1966 at the base of a shaft on the dry Nullarbor Plain, that fell in and died there about 4000 years ago. It would have lived in very different circumstances to the thylacines of Tasmania, by then separated from the mainland and affected in large areas by cold, moist Antarctic winds and cold ocean currents.

Predator follows prey (never the other way around!), and as the kangaroos and wallabies radiated into most parts of the continent, logic suggests the thylacine did likewise. Aboriginal people are known to have hunted them for food, but this would have been in a utilitarian manner, in no way threatening the species' existence. Instead, the thylacine may have been considered a prominent and respected part of the landscape. Ancient

Aboriginal engraved and painted rock art depicts thylacines in many styles and locations, scattered geographically and over time.

There are about 30 known thylacine fossil sites, out of a total of about 70 significant Quaternary mammal sites in Australia. In an evolutionary sense, therefore, there is evidence that the thylacine evolved as a successful marsupial, unchallenged as the continent's pre-eminent modern terrestrial carnivore.

3

AT THE END: EXTINCTION

> During our conversation, he told me that the last Tasmanian Tiger taken alive was caught at Adamsfield in the 1930s by his uncle. The method they used to trap the animal was to make a large box out of bush timber with a snare at the door and a baited release mechanism. The tiger had been around the camp for some time so it was not too long before it was caught in the box. To remove him from the trap they placed a rope noose, attached to the end of a long pole, around his neck. After that it was a simple job to hobble him and muzzle his mouth. Placed on the back of a horse he was delivered to Westbury Zoo where he was put on exhibition to the public.
>
> JOHN GOULDING, GEILSTON BAY

Extinction has always been integral to the life process, but when, as in the case of the thylacine, it is both manifestly unnatural and recent, it becomes our ineluctable duty to learn from the experience. Many have already done so. Alongside its positive allure and mystique as a tourism and quality-brand symbol, the loss of the thylacine glaringly symbolises wanton, careless destruction of the natural world.

Too late, Tasmanians realised that their champion marsupial carnivore occupied a vulnerable niche in the island's ecosystem. Thylacines were neither pest nor vermin but simply a perceived obstacle to nineteenth-century progress. And, whether snared or shot, they were also a source of bounty income.

Prior to European arrival, thylacine numbers probably averaged in the low thousands. Eric Guiler suggested a maximum of about 4000; Nick Mooney about half that number.[1] While the true figure can never be known, this is a reasonable estimate, based on predator/land-requirement studies. Tasmania is approximately the same size as Ireland or Sri Lanka, or West Virginia or Hokkaido, with about a quarter of its land surface permanently locked away as Tasmanian Wilderness World Heritage Area and protected parkland. Much of this area is inaccessible.

Five thousand is not a great number. What would it take to decimate this population in just one century, to the extent that it could not even find refuge in those wild areas of the state? Human predation, at a minimum, accounted for over two thousand thylacines presented for government bounty between 1888 and 1912. Many hundreds had been killed long before then, in response to a sheep-protecting private bounty and by hunting, already disrupting the population's stability. Habitat alteration simultaneously compacted the animal's range. On top of this, at the beginning of the twentieth century a virulent, possibly bacterial, infection may have fatally attacked many marsupial animals, including the thylacine. But there has been no conclusive evidence to support this as a contributing factor to its decline.

Naturalists, scientists and others, such as the artist John Gould, had been warning of the thylacine's possible demise since the middle of the nineteenth century. Such was the animal's lot, although alleged sightings continue, albeit increasingly rarely. (A 2017 Derwent Valley 'sighting' of a thylacine—a few seconds of grainy night video footage—attracted considerable publicity, but the animal was not a thylacine and was probably a spotted-tailed quoll.) The thylacine's continued survival would have to be a remarkable throw of the dice:

> Top carnivores, including eagles, tigers, and great white sharks, are predestined by their perch at the apex of the food web to be big in size and sparse in numbers. They live on such a small portion of life's available energy as always to skirt the edge of extinction, and they are the first to suffer when the ecosystem around them starts to erode.[2]

Furthermore, for the thylacine to still exist, it would have been successfully reproducing in sufficient numbers for many decades. This does not square with the reality of extinction. Two primary points are at issue. The first is that a species will become extinct if its numbers drop below the critical mass needed to sustain the gene pool, if the gene pool is limiting. The second is whether a species can survive once forced into a relatively difficult, increasingly shrinking wild environment. After millions of years of evolving as a specialised predator, what mechanisms would it have for 'adapting' to rapid, deleterious change?[3]

Scientists distinguish between mass extinctions and 'background' or 'random' extinctions, the latter occurring as a normal evolutionary feature, as first set out in Charles Darwin's 1859 *The Origin of Species*. The thylacine belongs in a third category, of human-induced extinctions, and so joins vast numbers of terrestrial, arboreal, aerial and aquatic species which have become extinct in just that way. (In this category, a northern hemisphere thylacine equivalent is the carnivorous sea mink of the rocky coasts of New England and Nova Scotia, hunted to extinction at the turn of the twentieth century.) There is however no evolutionary logic in ascribing positive Darwinian notions of survival of the fittest (that humans need to eat) to justify extinction of, say, the dodo or the passenger pigeon.

The statistics of human-induced extinctions can be mind-boggling. During the course of the nineteenth century something like 5 billion passenger pigeons were shot in the United States, along with up to 50 million Great Plains bison killed, and almost that many pronghorn antelope. Human ingenuity and cunning aside, many lost species simply lacked defence mechanisms against humans as predator. Dodos were not stupid: they simply had no inbuilt fear of people. This is a common trait among creatures endemic to remote islands where Europeans settled, such as Hawaii, New Zealand and Madagascar. Conversely, many creatures are naturally wary of man: an instinct developed over long periods of time.

Extinction is tricky: 'We have only the haziest idea about how many species currently exist on earth, how many there may have been in the past, and how many are going extinct at any one time'.[4]

The devastation of Australia's Ice Age megafauna was swift and severe. Thirteen genera of large marsupial mammals—about 45 megafaunal species—perished. This had occurred by about 20 000 years ago, when humans were well-established on the continent. Were they hunted to extinction? If they were, perhaps the thylacine survived because it was comparatively small and also mostly nocturnal. The thylacine in fact belongs in a rare subcategory of human-agency extinction, since it was exterminated not as a food (nor in the name of sport) but because it was supposedly destroying an economic asset, namely, sheep. Likewise the Falklands Island wolf, the warrah, was killed as a supposed threat to sheep and for its valuable fur.

Can random or background extinction explain the thylacine's disappearance from mainland Australia? This natural extinction process inevitably operates over a very long period of time and, equally inevitably, climate plays a role of some sort in it. Climate change alters growing conditions, which affect herbivores; in turn, carnivores are affected. It may be that the harshness resulting from continental Australia's aridification, and the sheer size of the continent, dispersed the thylacine gene pool to the point where introduced packs of hunting, scavenging dingoes more effectively took over the vast 'niche'.

Dingoes came to Australia with Aboriginal people, from regions to the north, not more than 12 000 years ago (when dingo-free Tasmania was cut off) and probably much more recently—about 6000 years ago.[5] Fossil evidence indicates that mainland thylacines were then widespread; the same evidence shows them to be extinct by about 3000 years ago. This rapid decline is of interest in itself, but is all the more intriguing for the fact that it parallels the decline of the Tasmanian devil on the mainland. Why did that strictly nocturnal, scavenging predator suffer a similar fate?

What is certain is that there have been two quite distinct thylacine extinctions, first on mainland Australia and New Guinea, and subsequently in Tasmania. There is no evidence that wild dogs competed successfully with thylacines in Tasmania after 1803—except in a cruelly oblique way, where thylacines were incorrectly blamed for dog-kills of sheep. More questionable is whether a competitor species, introduced through human agency, constitutes a 'natural' form of competition. Given the thousands of years over which the dingo–thylacine struggle apparently took place, one could say that struggle was natural. But the later introduction to Australia by settlers of many animals, which soon became feral, was not, setting in motion as it did endemic extinctions, destabilising an already fragile environment.

Those interlopers included mice, rats, rabbits, cats, horses, donkeys, camels, goats, pigs, water buffalo, foxes and cane toads, all, apparently, robust at adapting. The mammals are all placentals, needless to say. Does this prove the point that marsupials thrived in post-Gondwana Australia only because they had no placental competition? 'Perhaps' is not a very scientific word, but as an answer it will have to do. Contemporary feral destruction is, alas, not in doubt:

> The impact of introduced mammals on the Australian environment and fauna has been very great and is exceeded only by the case of New Zealand. It may reasonably be assumed, however, that each feral mammal species has by now spread into every suitable Australian habitat and that their populations are more or less stable.[6]

A stable feral population is proof positive that the ousting of some sensitive natives is complete. The long-term effect on the Australian biota of forced relocations or extinctions (the latter a growing list, headed by the thylacine) is not presently knowable, but there are sure to be adverse consequences. Sadly, the thylacine's demise may not represent an ecological catastrophe, or even a setback. We have lived without it. Yet if its passing means so little—and general indifference to threatened and endangered wildlife suggests so—the consequences for the future are truly grave.

A single species is estimated to have a lifespan of a few million years. In the course of natural events individual species die out slowly, over thousands of years, in the converse of the evolutionary process.

Set against all of the above, and rendering them puny, are the phenomena of the mass extinctions which have governed the fate of Earthly life and are but vaguely understood. There have been about fifteen of these events, of which five are considered major, and two so gargantuan as to be named the First Event (which came at the end of the Palaeozoic era, and destroyed 90 per cent of life on the planet) and the Second Event (the extinction of the dinosaurs). Some palaeontologists and other scientists believe that the Third Event has been underway for many thousands of years—demonstrably aided and abetted by *Homo sapiens sapiens*.

The Third Event involves the consequences of thousands of years of Ice Age change—in climates and sea levels—and the impact of human evolution. Four million years ago upright hominins were dispersing across Africa. Three million years ago, the famous Ethiopian, Lucy the bipedal *Australopithecus afarensis*, lived in a troop, her brain not much larger than a chimp's. Two million years ago *Homo habilis* appeared, with a brain size about midway between that of a chimp and the future *Homo sapiens sapiens*. In the last 300 years, the onset of the Industrial Revolution, and all that has followed, has greatly altered the natural world:

> Monitored wildlife populations—mammals, birds, amphibians, reptiles and fish—have seen a devastating 69 per cent drop on average since 1970, according to WWF's *Living Planet Report 2022*. The report highlights the stark outlook of the state of nature and urgently warns governments, businesses and the public to take transformative action to reverse the destruction of biodiversity.[7]

Increased anthropogenic carbon emissions have created global warming out of step with nature's processes for doing the same. The Earth's lungs, the tropical moist forests, are

The thylacine, along with the dodo, is the most prominent victim of modern human-induced species extinction. But unlike the flightless bird of Mauritius, which became an easy food source for Dutch sailors, the thylacine was targeted as an economy-wrecking sheep killer. Painting attributed to Dutch artist Roelandt Savery, c. 1626. The Natural History Museum's first superintendent, Professor Richard Owen, used this painting to scientifically describe the bird.

themselves headed towards extinction. One does not have to be a Third Eventist to be awestruck at our capacity to destroy without replenishing.

Meanwhile, it is no coincidence that Tasmania, the island that killed its tiger and has regretted it ever since, has much of its land mass locked away as World Heritage Area and parkland. This makes it one of the world's most protected places. Few would dispute the role of the thylacine in making that come about, as will be discussed in a later chapter.

The thylacine may be officially long gone, but other 'extinct' animals have come back and, through global agencies such as the

IUCN (International Union for Conservation of Nature) and CITES (the Convention on International Trade in Endangered Species), some imminent extinctions are being arrested—although sadly these tend to be the high-profile cases such as the panda and Russia's 40 known Siberian leopards.

Extinction reversal, as a natural return from the dead, is invariably surprising. The coelacanth, a deep-sea fish, was believed to have been extinct for 75 million years until discovered alive and well near the Comoros Islands off Africa's east coast. It is unchanged in form, a Palaeozoic miracle. Ironically, nearby Madagascar has been stripped of its forest cover and has possibly the inhabited world's worst extinction record. Yet in the 1980s the golden hapalemur was discovered there, a new species of lemurid. Numerous Australian marsupials and birds have been declared extinct only to be found again. The noisy scrub bird is a good example, last seen in 1899 before its rediscovery in 1961. Other returnees include the bridled nail-tailed wallaby and the sandhill dunnart and, for sheer longevity of absence, the mountain pygmy possum, believed extinct for 15 000 years.

The Angolan giant sable antelope, thought to have been rendered extinct by 30 years of sustained, brutal civil war, has been rediscovered, with a few hundred surviving in a national park, as Angola's National Animal. Angola is much larger than Tasmania. They do, however, have in common great tracts of forest which people seldom enter. No doubt the antelope retreated there. It is this ability of a species to protect itself that gives so much hope to those who believe in the continued existence of the thylacine.

A detailed University of Tasmania study released in 2023 identified Tasmania's remote south-west as a potential refuge for a remnant thylacine population. A research team led by environmental sustainability expert Professor Barry Brook analysed over

1000 observational records dating back to the early 1900s. They showed that the thylacine's distribution shrank rapidly because of the bounty schemes. While that is not surprising, the research led to an interesting conclusion. In the words of Professor Brook: 'We found, through further analysis, that extinction might have been as recent as the late 1980s to early 2000s, with a very small chance that [the species] still persists in the remote south-western wilderness areas.'[8]

Those large areas, protected as World Heritage–listed wilderness, and virtually free of human activity, might appear therefore to be an ideal refuge. Set against that optimism is the reality that they are not suitable thylacine habitats, having non-nutritional vegetation resulting in low animal density and therefore constant food scarcity for a carnivore. That being so—if thylacines are not living undetected there—the words of University of Sydney academics Peter Banks and Dieter Hochuli are a stark reminder of what really happened. In *Thylacine: The History, Ecology and Loss of the Tasmanian Tiger* they write: 'The thylacine was hunted to such low numbers, and its environment changed so much, it was sucked into an extinction vortex.'[9]

4

'PATHETICALLY LITTLE IS KNOWN'

> It seems lunchtime had arrived at Adamsfield and Mr. Batiste and his partner were seated at the sheltered edge of a clearing, where they were working a sluice used to recover the alluvial osmiridium. In Mr. Batiste's own words, or as near as it is possible to get to them: 'We heard a commotion in the bush behind, when a wallaby burst out right in front of us, bounding and weaving for dear life across the open space, closely followed by "His Nibs" in full cry . . .'
>
> GEOFF ASCHMAN, LINDISFARNE

What, then, *is* a thylacine? Until officially declared extinct on 7 September 1986, it had the distinction of being the world's largest extant marsupial carnivore, existing only in Tasmania, the cool-temperate island state of Australia. Long part of the mainland, Tasmania became separated from the rest of the continent by a body of seawater, Bass Strait, some 12 000 years ago. That interglacial rise in sea-levels, a result of global warming, saved the thylacine, which was headed for extinction on the mainland.

Although the thylacine has always been an intriguing and controversial subject and therefore much written about, it wasn't until 1985 that the first definitive book on the animal was published, Eric Guiler's *Thylacine: The Tragedy of the Tasmanian Tiger.* It was then, and still is, a critically important account of the animal. The title of this chapter is taken from it:

> Pathetically little is known of the biology of the thylacine. It is possible to make reasonable deductions from the early anatomical descriptions but we have to complete the picture from the other sketchy information available. However, this is not made easy for us as the literature contains some contradictory statements and some observations are obviously incorrect.[1]

Robert Paddle's *The Last Tasmanian Tiger: The History and Extinction of the Thylacine*, published in 2000 and arising out of the author's PhD thesis, was a timely and meritorious new contribution to the understanding of the thylacine, particularly in respect of attitudes towards the animal. The book is meticulously researched and provocative. Where Guiler cites a lack of information as the main difficulty in thylacine research, for Paddle the real stumbling block has been the inconsistent and sometimes deliberately false information attending the animal before and since its demise: 'There are ambiguities, silences, real omissions, pretended omissions and contradictions in the literature.'[2] Thus, in considering the thylacine's anatomical and behavioural characteristics, it is as well to bear in mind that fact and folklore are by definition at odds.

The 2023 CSIRO publication *Thylacine: The History, Ecology and Loss of the Tasmanian Tiger*, edited by Branden Holmes and Gareth Linnard, provides excellent new and re-assessed thylacine material, having no less than 78 contributors working in many disciplines internationally.

In respect of the thylacine's biology, it is as well to begin at the beginning: the thylacine was a marsupial mammal, that is, it

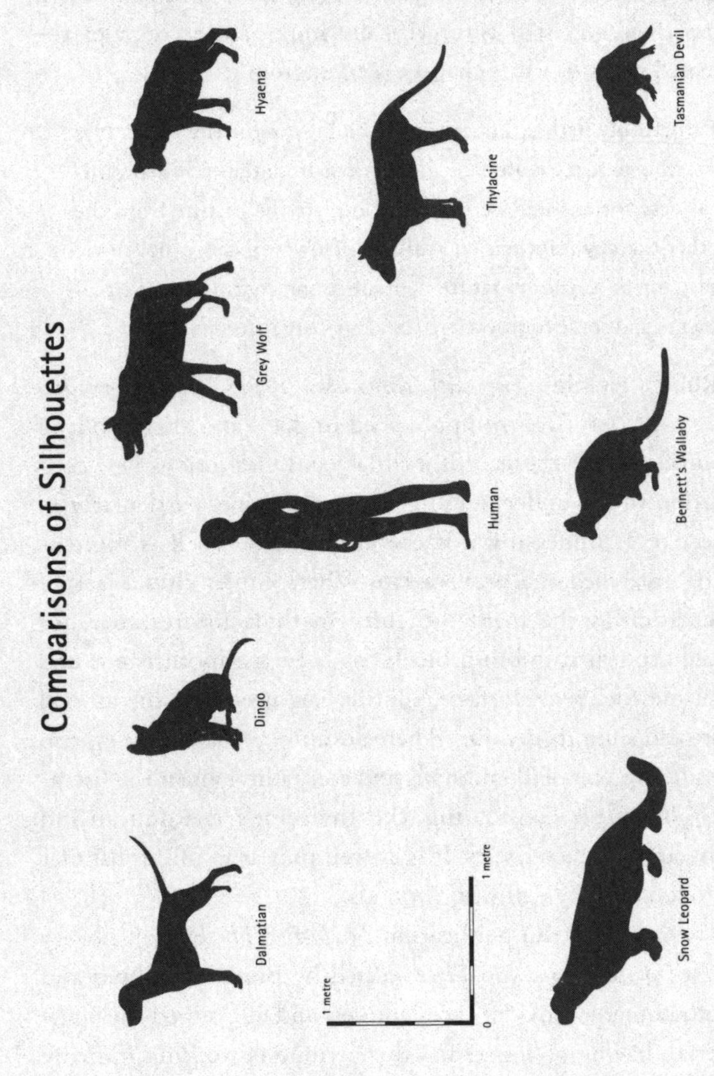
Comparisons of Silhouettes
Dalmatian
Dingo
Grey Wolf
Hyaena
Human
Thylacine
1 metre
0
1 metre
Snow Leopard
Bennett's Wallaby
Tasmanian Devil

gave birth to tiny immature embryos suckled in a *marsupium*, or pouch. (Marsupials inhabit the Australian continent, which includes New Guinea and some nearby islands, as well as North and South America.) Thylacines were short-haired and tawny or tan-coloured. Their most distinctive feature was the sixteen or so dark stripes running across the back and sides, in various lengths and widths, from the base of the shoulder onto the base of the tail.

What size were they? Museum specimens give a fair indication, and one can speculate that some of the older written accounts were exaggerated. A report that may be presumed credible, from a newspaper in 1885, concerns a shepherd and his dog near Launceston investigating a sheep kill, one of five over an unknown period of time. The dog got into 'a severe fight [with] a very large tiger'. The shepherd shot the tiger, 'which measured 5ft. 4in. from the nose to the tip of the tail'.[3] If one assumed a mean adult body weight of about 25 kilograms, that animal would equate to the size of an average adult striped hyaena, African wild dog, snow leopard or Dalmatian.

The illustrations on page 38 provide useful mammal size comparisons, although it is important to remember that existing photographs and moving images of thylacines show considerable size variations between adult males and females, males being considerably larger.

Although the thylacine's head shows a distinct canine profile and alertness, that resemblance is less so at the bulging jawline. The famously wide gape of the jaws is a strong indicator of its predatory adaptation. A vice-like grip, preferably at the head, neck or upper body region, brings the prey to a halt and kills it by suffocation and/or internal crushing. Writing in the Holmes and Linnard publication, academics Marie Attard, University of London, and Stephen Wroe, University of New England, observe that: 'Thylacine skulls have a long snout that would enable them to bite down at high speed, a characteristic of predators specialising in relatively small, more agile prey and those that use a pounce-pursuit or ambush hunting style.'[4] (To the plethora of misplaced other-animal names given to the thylacine, should we now add 'Van Diemen's Land crocodile-wolf'?)

Analytical comparison with the wolf does remain valid. University of Melbourne evolutionary biologist Axel Newton and colleagues carried out thylacine–wolf skull shape comparisons, utilising a technique known as 3D geometric morphometrics: 'We found that not only do the adult thylacine and wolf skull share superficial similarity, but they also share greater statistical similarity with each other than the thylacine does with its own closest living relatives, such as the Tasmanian devil or eastern quoll . . . This was a truly unexpected finding, given that the thylacine and wolf last shared a common ancestor over 160 million years ago.'[5] (It should be remembered too that the gape was multifunctional, the social threat-yawn function of the thylacine gape being similar to that of the Tasmanian devil.)

Captive thylacines were seen to consume some bone as well as the flesh of carcasses given to them. However, their wide jaw gape does not equate to a bone-crunching ability; the size and distribution of their teeth do not support that notion. Their once-populous relatives, Tasmanian devils, on the other hand, have immensely strong jaws and will readily consume bone.

Studies of the thylacine's musculature and skeleton also engender debate. It possibly lacked the ability to stiffen the wrists and digits, which is a requirement for quadrupeds to run at speed. This would lend weight to accounts of its being a ponderous or awkward-looking runner. A presumably authoritative 1983 Australian Museum account stated: 'It had a rather stiff gait and could not run very fast'.[6]

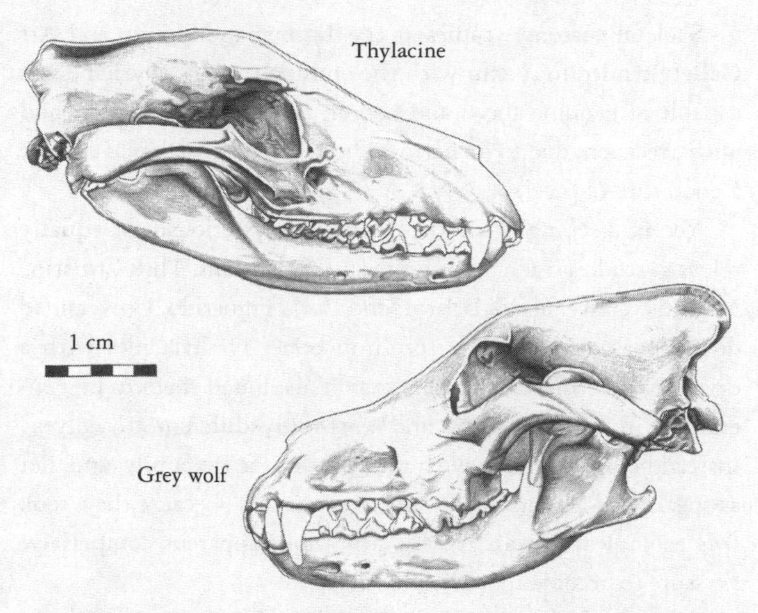

A comparison between the skulls of the thylacine and the grey wolf. Note the thylacine's massive molars at the back of the jaw.

Paddle cites evidence to the contrary, and also refers to their 'rapid pursuit' hunting technique.[7] Academic Natalie Warburton, Western Australian Museum and Murdoch University, reaches a similar conclusion. Her skeletal study 'suggests that, in comparison with all other marsupials, thylacines were well-adapted for running quickly and efficiently, presumably in order to catch prey.' She continues:

> Long, thin limb bones . . . enable longer strides, such that distance can be covered more efficiently than with short limbs. The muscle attachments on the long bones correspondingly reflect a relatively 'high-geared' arrangement, whereby muscles are able to achieve a distance or speed advantage. Long in-levers such as the calcaneus (heel bone) similarly suggest investment for rapid limb movements . . .[8]

Skeletal anatomy studies at the Tasmanian Museum and Art Gallery tended to concur with the opinion that the thylacine was capable of genuine speed and had an ability to turn swiftly and with precision, due to its narrow chest. (Fleeing wallabies exhibit a good turn of pace!)

Yet in keeping with the animal's mysteriousness, equally relevant studies reach other tentative conclusions. Thus Christine M. Janis, University of Bristol, and Borja Figuerido, Universidad de Málaga, in comparing forelimb bones of thylacines with a diversity of living carnivorous mammals, found them to be generalised in their anatomy and 'distinctly different to wolves', instead being 'clustered with members of the cat family' and 'not as specialised for running as placental cursors', because they 'took this ecological role in Australia without apparent competitive pressure to become more specialised'.[9]

Slowing the species down even more, these presumed techniques must in turn contend with contradictory evidence. Well-known Tasmanian bushman-trapper Mont Turner in the 1960s spoke of them 'prowling along patiently after their prey'.[10] (As a child he had helped his father trap the two thylacines which are in Launceston's Queen Victoria Museum and Art Gallery.) And there are many other references to their slow pursuit technique. Yet of all the contradictions, none is surely so incorrect as the frequent references to the thylacine having a stiff, awkward tail that made turning difficult. Why would nature spend millennia endowing a predator with such a cumbersome appendage?

Earlier skeletal studies had been used to argue in favour of the animal having a bipedal tendency, that is, an ability to hop like its distant relatives the herbivorous marsupials. Certainly, the rear leg could be flattened at the ankle, but to thereby infer kangaroo-like locomotion requires a metaphorical leap of faith and imagination. The thylacine's far closer marsupial relatives,

Dr David Fleay's photograph of a thylacine at Beaumaris Zoo, December 1933. The image is a classic example of the threat-yawn display used by Tasmanian tigers and Tasmanian devils when agitated, excited or challenged. (Dr David Fleay, courtesy David Fleay Trustees)

the Tasmanian devil and the tiger cat, are strictly quadrupedal. The belief in a bipedal form of movement may have arisen in response to a number of early written accounts that refer to it briefly bounding and/or hopping, particularly when chasing or being chased.[11]

Cat-like abilities were also ascribed to it, in being able to jump fairly high and also to leap with agility from one raised point to another. Exaggerated or not, all of these features would require a sound balancing mechanism. The thylacine had a long, powerful tail; surely a prominent asset, though far from having a truly pentapedal (fifth-limb) function. The tail could be used as a rudder, to improve the animal's balance and thus its ability to run and turn swiftly. The tail also had a clear social function: swishing/wagging to indicate anger; held out straight when disturbed or threatened; and held vertically when sexually aroused, for both visual and olfactory communication. A crest of hair at the end of the tail could be raised, like hackles (this is also a feature of the Tasmanian devil).

The following 1899 account, if accurate, vividly describes the tail—and is a rare instance of a person being attacked:

> Adventure with a Native Tiger.—It is by no means generally believed that there are native animals that will face a man. Nevertheless, it is so. On Monday, as Mr. B. Stevenson, a sheep farmer on the North Esk, was going round his run, a large native tiger rushed close up to him. He hit it with his walking stick, which broke with the blow. The tiger, uninjured, turned and ran away about ten paces, then faced round, growling very fiercely, and came up a step at a time to within 6ft. or 8ft. Then he stood growling, and with his tail wagging backward and forward after the fashion of a cat on the point of catching a bird. Mr. Stevenson's only means of defence was his pocket-knife. He could not even get hold of a stone. However, after due consideration the animal evidently deemed discretion the better part of valour, and made off for an adjacent dogwood scrub. Formerly tigers were very troublesome to the sheep farmers on the North Esk River.[12]

Finding evidence of the thylacine's contact with the ground in the form of spoor (paw-prints), finding scats, or a lair, have long been held out as the most likely proof of its continuing existence. Its prints are not dissimilar to a dog's and, as previously noted, might also be mistaken for a wombat's. A dog-like quality also applies to its droppings, which again might bring such evidence into question. A typical lair was known to be a shallow rock cave with nesting material covering the floor. Hollow logs, dense shrubs, overhangs and rocky outcrops were also used. A captive thylacine at Washington's Smithsonian Institute zoo was observed to carry hay in its mouth for nest-building. It is also possible that, like devils, thylacine families used more than one lair. (Hyaenas frequently move

A unique animal. As an 'extension' of the body, the tail is powerful and possibly rudder-like; the hind legs appear to be shorter than the forelegs but this is because the versatile marsupial ankle flattens, creating a 'foot' which might have had a residual hopping function; the thick neck supports an extraordinary jaw. (Michael Sharland)

their cubs between lairs.) Whatever the case, discovery of a lair in current use in the extensive Tasmanian wilderness would cause great excitement, and would be surpassed only by an unambiguous sighting.

Any number of colourful and imaginative descriptions were applied to the noises the thylacine made. Writing down a sound is difficult at the best of times, which partly explains the uncertainty over this particular attribute. Needless to say there are no recordings of its vocalisations. Confusion with other animal noises also played its part, most notably with the shrieks and screams of the Tasmanian devil at night—noises which are most unsettling to the uninitiated. A 'coughing bark' would appear to describe the thylacine's more usual vocalisation; also, a 'rapidly repeated kind of cough'.[13] An expert who would have heard it many times, Alison Reid, Hobart Zoo's last curator, wrote the sound as 'ah-ah-ah-ah'. This indicates a muted level.

In describing his thylacine capture, Mont Turner recollected that the animal 'couldn't get free and he coughed with rage. It was the same cough I'd heard many a night as I lay in my bunk in the camp'.[14]

But this was just one part of the vocal range. Paddle devotes considerable space to reconstructing and analysing five distinct vocalisations and provides written evidence that the sounds could be quite loud. Far less inclined to be certain—reflecting that uncertainty is the governing factor in so much information about the thylacine—is David Pemberton. He cites a woman who reported that it has a 'ping'-sounding call. Were that to find its way into the media, says Pemberton, 'then suddenly we have the sixth call of the thylacine: the ping call'.[15]

Annis Hardcastle Knight, the daughter of famed US animal artist Charles R. Knight, collaborated with him in describing the thylacines at the Washington, DC, National Zoo. In 1902 she wrote: The call is a peculiar coughing sound, which, until recognised as a note of affection, gave the keepers grave concerns regarding the condition of her lungs. The sound was easily understood when it was observed that the mother, in giving vent to it, was immediately answered by her cubs. Often when sleeping she will rouse for a moment, and, after making this peculiar cry, listen expectantly until she hears the diminutive counterpart, whereupon she will let fall her head and rest content.'[16] Across the Atlantic, the Secretary of the London Zoological Society stated that 'it makes a prolonged, very loud, undulating cry . . .' By any reasonable inference that was sorrow.[17]

Why the stripes? They too were surely multifunctional. The traditional explanation is camouflage but, as with the stripes of the zebra, reassessment of the seemingly obvious is no bad thing. Individual recognition was a likely function, likewise the converse of camouflage: a visible predator striking sudden panic into a prey group, which flees, with the weakest becoming the target. Striped and spotted cats hunt like this.

A less desirable outcome of having a striped hide is that a thylacine waistcoat allegedly became a sought-after fashion item in Victorian England and in the United States. Although folklore has it that they were popular, no such waistcoat is known to exist, and the Tasmanian Museum and Art Gallery's Kathryn Medlock, who developed the museum's thylacine collection and researched thylacine collections and information worldwide, is sceptical: 'It was written once by [thylacine fanatic and former head of the Tasmanian Film Corporation] Norm Laird and it's been quoted ever since. I've been through his papers and I've been through export deeds and we've never found anything. Apparently one was sold to the Victoria and Albert Museum, so I wrote to them and they said no, that never happened'.[18]

Reproduction and its consequences arc across anatomy and physiology. The female thylacine raised up to four young, having that number of nipples in her backward-opening pouch. Breeding was thought to take place every two or three years. It follows that the thylacine family unit was small. (African hunting dogs have up to sixteen pups in a litter.) What is certain is that the young stayed permanently in the pouch for the first few months of life as fur and limbs developed, then spent a period of time outside the pouch and fully mobile, but still returning to suckle until weaned. Thereafter, speculation and conflicting hearsay again take over. Did the cubs run with the mother while she hunted, or did she stash them in the den, returning there with food for them? The relatively large number of mothers and cubs captured for zoos and bounty claims indicates that the cubs were often with the mother. (Mary Roberts,

Thylacine males and their closest relatives Tasmanian devils have pseudo-pouches in which the testes are normally protected. (Queen Victoria Museum and Art Gallery)

who owned and ran the original Beaumaris Zoo in Hobart, wrote that thylacines were attentive mothers.)

Dr Heinz Moeller, a thylacine expert and career director at the Zoological Institute of Heidelberg University, in 1997 published a significant academic work, *Der Beutelwolf: Thylacinus cyncocephalus*. In this work Moeller describes the morphology of the thylacine brain, including that its well-developed cortex area may be associated with complex social behaviours. Similar findings are published in the CSIRO's *Thylacine: The History, Ecology and Loss of the Tasmanian Tiger*. Contributors Gregory S. Berns (Emory University, Atlanta, US) and Ken W. S. Ashwell (University of New South Wales) write that the thylacine cortex is proportionately enlarged when compared with that of the Tasmanian devil, the bigger size being associated with complex cognition. They state that, 'this would be consistent with the thylacine's ecological niche as a predator, which would require more planning than the scavenging strategy of the devil.'[19]

An unlikely historical source corroborates these contemporary findings. George Augustus Robinson was an evangelist appointed in 1829 by Lieutenant-Governor George Arthur to travel the island and 'conciliate' the 'hostile' Aboriginal tribes. His attempts to pacify the Indigenous peoples into accepting British civilisation were well-meaning but had a deeply flawed outcome. Nonetheless, the copious diary writings Robinson kept while travelling the island and living mostly amicably with Aboriginal people for five years are a unique and extensive record of their customs and languages. And, incidental to their central purpose, his diaries contain many thylacine observations. The diaries were edited and published in 1966 entitled *Friendly Mission*; a second edition published in 2008 is a hefty 1162 pages, from which the thylacine descriptions reproduced in Chapter 7 of this book are extrapolated.

A number of Robinson's thylacine observations describe the mother and her pups running together, feeding together, socialising and hunting together—distinctive proof of complex cortex-derived activity. Thylacine family groups never included an adult male. That being so, an intriguing further possibility relates to instinctive and even learned protection behaviours against scavenging Tasmanian devils. This would require thylacine cohorts to overlap, that is, juveniles still at foot with the mother, with the subsequent litter helpless in the den and subject to predation while the mother was away hunting. The juveniles might well have stayed on effective guard duty at the den at such times.

The fundamental notions of territory and home range must also be inferred, because conflicting first-hand reports indicate the animal as having been strongly territorial but also as wide-ranging and even nomadic. (The Tasmanian devil is not territorial.) Could it be that as the European settlers dispersed across the island state they observed both natural thylacine behaviour and behaviour influenced by human encroachment? Guiler suggests that the killing of thylacines freed up choice areas of territory into which new individuals or family groups moved, so creating 'an impression of constant movement'.[20]

It is likely that in their natural state, perhaps especially restricted to an island, with distinct microclimates, thylacine populations would tend to confine themselves to defined, scent-marked ranges. Thus Guiler's view makes sense—that when nature's equilibrium is disturbed, unlikely patterns emerge and can appear to be normal.

There remain two aspects of thylacine behaviour to consider. One is the animal's diet, the other its relationship to humans. Its speed and agility—or lack of either or both—would clearly have a bearing on how it hunted. Thylacines are believed to have overwhelmingly targeted macropods (wallabies and kangaroos) as prey. Adult kangaroos are considerably larger than thylacines. These prey animals, which are both reasonably swift, are active from dusk to dawn, generally feeding on open plains beyond the treeline. This suggests an ideal scenario for the cooperative hunting seen on the African plains, in which the prey group is panicked and flees and a weak individual chased and brought down. Thylacines were reported hunting both singly and in small, family-sized groups. Cooperative techniques would appear feasible for an animal that wasn't fast (beyond a short sprint), that pursued prey sometimes big or bigger than itself, and that had a known ability to pursue until the prey was exhausted.

The following observation by animal behaviourists Jane and Hugo van Lawick-Goodall of East African hunting dogs (an endangered species), which they call nomads of the plains, can with some justification be imaginatively superimposed onto a marsupial meadow of the pre-European island, since it describes a technique which the thylacine is known to use:

> Wild dogs normally approach a selected individual or herd very slowly, walking with their heads held low and parallel with the ground, and adopting a slight crouch in their gait. In this way the hunters can sometimes get to within fifty yards or so of a herd of zebras or wildebeests before the intended prey begins to run . . . From this point, the hunt may develop in a number of different ways. Sometimes, particularly when the dogs are approaching a small herd, it appears that the prey is selected before the chase begins, often by the leader of the pack. When he starts to run the other dogs follow suit, and all dogs pursue the same prey until it is either caught or manages to escape. At other times, usually when the wild dogs are hunting large herds, the pack makes a short chase towards a group of animals and then stands or walks slowly whilst watching intently as the herd starts to run . . . What are the dogs looking for when they stand watching a herd run past them? And why do the individuals of a normally closely united pack sometimes separate in seeming disorder? [Because] both techniques enable the wild dogs to select an individual from the herd that is, in some way, weaker and slower . . . During the actual chase wild dogs placed behind the leaders will cut corners when the prey changes direction and so gain positions farther forward in the pack, or actually in the lead. This is particularly noticeable when the dogs chase a Thompson's gazelle, for this creature usually zigzags across the plains when hunted, or describes a very large circle, so that, by the end of the chase, several different dogs may have led the pack simply by cutting the corners.[21]

A wallaby when hunted also flees in a wide circle, which the thylacine was observed to cut across, finally ambushing it after what may have been an exhaustive chase. While Guiler notes a lack of evidence to support pack hunting, Paddle, on the other hand, sees 'placental chauvinism' at work in falsely condemning the thylacine to a lack of speed and therefore to a hunting technique relying solely upon dogged pursuit. He suggests that dynamic thylacine striping enabled individuals to instantly recognise each other: a vital part of daytime cooperative hunting. African hunting dogs are also vividly, idiosyncratically marked.[22]

So it is that the stripes possibly had a hunting function. Can the same be said for the tail? Arising out of a number of early assumptions, the thylacine was credited with being a marine predator, and a good one at that. Settlers had learned from the island's Aboriginal people that thylacines were strong swimmers. This in itself is a common mammalian characteristic, as most mammals have bodies of water as part of their habitat. But from the very beginning scientists reckoned the compressed–flattened–stiffened tail to be adapted for swimming, and this, combined with obvious empirical evidence—the early white settlers were all on the coast—of thylacines roaming the seashore, led to the not unreasonable belief that they ate crustaceans, seal pups and washed-up marine carcasses, and even chased fish.

The earliest description of the Tasmanian devil accurately claimed that it ate blubber. This is a known devil scavenging feature. While evidence of the thylacine as an opportunistic scavenger is not abundant, a lengthy quasi-scholarly article written in 1900 is informative. And truly wild! A group of six men were sailing in the Port Davey region of south-west Tasmania prospecting for gold and silver, when they saw a large whale carcass, which they decided to row their tender boat to and cut out the whale's ambergris. But they hadn't reckoned for a scene rivalling Alfred Hitchcock's 1963 natural horror thriller *The Birds* (and it reads at times like fiction):

As we rowed in to the beach, we saw that there were skulking along it three or four of those large yellow, black-banded wolves, sometimes called Tasmanian tigers, and we saw two smaller black and white wild cats [probably devils]. The whale lay just awash in the gentle surf. A cloud of gulls and other sea-birds were hovering over the place, but not feeding, although evidently very hungry and in constant motion. Perched on the carcase were seventy-five or a hundred larger birds, which we thought were eagles [they would have been Pacific gulls]. Such an assemblage of eagles none of the party had ever seen or heard of before . . . They were monarchs of the feast, too; for the striped 'tigers' stood at a distance, or approached very stealthily for a share of the blubber . . . We landed fifty or sixty yards from the whale, and Devoe, with the fowling-piece in his hands, stole forward in the hope of getting a shot at one of the striped wolves. They all made off, however, before he could come near enough to injure them with buck-shot. He fired at one of the black-and-white cats; and with the report and the squallings of the wounded beast, the eagles rose with a mighty flapping of their wings and a chorus of hoarse screams . . . One swooped down, and with a harsh, whistling scream, passed so near the head of the Auckland engineer as to brush off his hat . . . One would no sooner brush past than another was descending like a dart. And now, excited by the screams of their fellows, a cloud of eagles were on the wing from the cliffs . . . Cathcart was knocked off his feet and nearly blinded by another eagle, which struck him heavily from behind . . . the rushing noise made by their wings was so bewildering that we were not a little upset . . . By this time his face and hands were painfully torn . . . I still swung the gun butt, and contrived to inflict some little injury on our winged assailants, but was constantly buffeted and received a number of ugly scratches . . . As the conflict

> proceeded, they appeared to scream more wildly and become more incensed . . . The skipper received an ugly talon wound on the neck . . . All five of us were now [hiding] under the [upturned] boat, and the exultant eagles circled round, screaming in triumph . . . All five of us were torn painfully . . . Gradually the warlike birds made off to alight on the cliffs, or to return to their repast on the whale; but after being cooped under the boat for nearly an hour we seized a chance to leap out, shove it off from the beach, and row hastily out to the sailboat. A dozen eagles rose and came screaming after us, but they did not renew the attack in force.[23]

The eminent naturalist Michael Sharland, writing in the 1960s, stated as fact that 'being essentially a hunter and a discriminating one, satisfied only with warm flesh, it was natural for it to prey upon sheep and lambs which were easier to hunt than native game'.[24]

Generating untold harm were the myth-making, demonising assertions relating to the thylacine's alleged blood-thirst. Experts do not disagree on its predilection for vascular tissue and some of the internal organs, such as the kidneys, heart and lungs. Not only do these organs provide excellent nourishment for a carnivore; in the natural state on the island the thylacine had no quadruped competitor obliging it to eat every last morsel of prey on the spot, or to drag the carcass to a place of safety. Graphic colour photographs show allegedly recent, selective-feeding thylacine kills.[25]

It is only a small step leading from selective, nutritious feeding to the tag of vampire and all the old European-based myths of lurking horrors in the dark woods. Where contemporary opinion diverges is on the extent to which the thylacine actually lapped or even 'drank' blood. Paddle constructs a case to lay the blame for this notion squarely on the shoulders of a snobbish Oxford University academic who visited Tasmania in 1907, and whose ill-informed views were subsequently swallowed by most Australian scientists, including Guiler apparently, who as recently as 1991 wrote: 'Tasmanian tigers would kill sheep in a very characteristic way. They used their huge gape to bite out the throat and then they drank the blood'.[26]

Or, as one populist account would have it: 'Many believe it was not so much the flesh that the tigers killed for but the blood which they sucked through the neck and liver',[27] suggesting a vampire-like carnivore that had little or no need of solids! (The author does, however, also state that it ate 'almost any other animal or bird it could find in the bush'.) Kathryn Medlock expressed concern that not only is the vampire tag wrong, it is also of recent origin. Former University of Tasmania senior lecturer in Zoology Dr Randy Rose is another of the opinion that, as a carnivore, the thylacine would not be too fussy a feeder.

Captive behaviour is no substitute for natural behaviour. But this personal 1863 observation may have merit: 'Some idea of her extraordinary muscular power may be formed from the fact, that since her capture she has frequently been supplied with live kangaroos of every size, and has in each case killed with as little difficulty as a cat would find in killing a mouse.'[28] And if this 1905 account is accurate, thylacines weren't beyond opportunistic scavenging:

> Concerning the delay of the reporting the finding of the wreck of the *Acacia* before the wreck of the *Brier Holme* [near Port Davey, loss of 17 lives], the fishermen concerned say that bad weather prevented communication. Their narrative concerning the finding of the denuded skeletons is almost similar to that concerning the *Brier Holme*. The bodies had been gnawed by

> Tasmanian tigers, one of which was found dead near the corpses, having apparently been bitten by snakes, many of which were found near the bodies, and were killed by the party.[29]

Alas, it seems to be true that we know pathetically little. And yet many Tasmanians interacted directly with thylacines and were, mostly, honest in telling their stories. It is from them that other glimpses of the genuine creature can be seen. (In a double irony, such is its mystique that even today some Tasmanians know of someone who knew someone who was bushwalking at dusk when suddenly . . . and the myths and imponderables begin all over again.)

Naturalist and journalist Col Bailey collected 'tiger' stories for many decades and his 2001 publication *Tiger Tales* provided a detailed and intensely researched perspective on the thylacine. Bailey's first-person accounts of wild thylacines and their behaviour are important as an alternative to their demeanour in captivity; not least when the general impression, as befitted the times, seemed to be that zoo thylacines were boring and smelly.

Bailey befriended an old Depression-era trapper by the name of Reg Trigg who, back then, had caught a 'terrified' thylacine in a snare. In time Trigg was able to feed it by hand and stroke the top of its head, but he eventually released it back into the wild.

Pet thylacines were not all that uncommon but could probably only be domesticated when caught as pups. They could apparently be walked on a lead and were generally docile and even friendly. Botanist and Royal Society fellow Ronald Campbell Gunn, who in 1842 became editor of the *Tasmanian Journal of Natural Science*, tamed no less than three and, presumably accurately, wrote: 'It seems far from being a vicious animal at its worst, and the name Tiger or Hyaena gives a most unjust idea of its fierceness.'[30]

In an 1885 issue of *Field* magazine, William Bernhardt Tegetmeier recorded an interesting behavioural account of thylacines at London's Zoological Gardens: 'After feeding, they will play very much in the same manner as two dogs, opening their enormous mouths and pretending to bite each other. They are so far domesticated that they permit the keeper to go into the cage, and will even rub against him.'[31]

Tegetmeier also included this quote from Ronald Gunn, who had sent a female thylacine with three pups to London's Zoological gardens in 1863:

> The [mother], in giving suck to its young, used to lie down like a dog, the skin of the pouch being thrown back so as to admit of the young ones getting easily at the teats. When alarmed, the young ones crawled in with their backs downwards, the mother assisting by lowering her hind quarters to facilitate them getting in, and by also placing her rump against the side of the cage to give the cubs a purchase with their hind legs against the cage, and thus push themselves in. They were so large when they left this, that when they were all in the pouch it hung down very low, and seemed almost a deformity.[32]

(Tegetmeier also noted a decidedly inaccurate opinion formed by British naturalist William Swainson: 'The tail is so singularly compressed that Swainson, whose zoology was of a very ideal character, concluded the animal was aquatic.'[33])

Despite their general elusiveness and retiring nature, numerous stories relate to thylacines in the wild being naturally inquisitive about humans and following them persistently—by one account, for nearly two days. Thylacines hanging around campsites is more explicable, given the presence of food. And in the natural state thylacines and Aboriginal people coexisted for millennia.

The pity is that those who were interested in the thylacine represented a very significant minority, as epitomised in Bailey's account of teenager Stanley Conroy, who in the early 1900s was working with his father clearing bush. One morning the youngster

> was astonished to see a fully grown male Tasmanian tiger casually inspecting the ranks of the yoked bullocks. Stanley stood mesmerised as the tiger boldly strolled among the terrified oxen. Once the animal caught wind of the human . . . however, it vanished in an instant into thick bracken lining the track. Stanley kept the sighting to himself for he knew that should his father learn of the tiger, he would not rest until he had hunted it down and shot it, such was his hatred of the animal.[34]

Still, not every rural Tasmanian harboured such irrational negative passion: 'If anyone would like to see a specimen or two he should go out to the run at night, take a hurricane lamp and stand it on high ground in a clearing, and sit on the leeward side. Any of Tasmania's wild animals will be attracted by a light, which has only to catch the hyena's eye when he will creep carefully to within about 20 yards of it.'[35] That seasoned observation was made in 1937, the year after the death of the last known thylacine.

5

A RUGGED AND DETERMINED FRONT: VAN DIEMEN'S LAND SETTLED

About half way between the Frankland and Arthur Rivers we were camped for the night, in a beautiful glade in a myrtle forest, when around the fire in the evening, my father told of the experience he and Bill Morley had as they walked into Mount Balfour to commence work at the mine, over the same track, and he told us that as they walked in they were stalked through the night by two Tasmanian tigers, and were afraid to stop for rest . . .

R. R. McArthur, Lindisfarne

Over a period of some thousands of years a profound technological development set up conditions for civilisation to advance. This was the domestication of plants and animals, which first occurred in the Mesopotamian region. Where previously wild grains had been harvested where they grew naturally, between about 8000 BC and 3000 BC barley, oats, lentils, olives, onions, camels, cattle, sheep, pigs and horses all came under the control of man. The invention of irrigation and the plough and the warming global climate made cultivation possible. Far eastwards across the other side of the world, at the south-east tip of Australia, that same global warming had caused the Bassian Plain to be flooded, creating the strait that isolated Tasmania from the mainland.

Thus, while the peoples of the Fertile Crescent were establishing the first villages and working out rudimentary concepts of commerce, maths and philosophy, Aboriginal tribes became isolated on a large island at latitude 42 degrees south. People had already been in those parts for tens of thousands of years and their ancestors had experienced similar comings and goings on the low-lying flat plain, bounded by mountains to its east (their remnants now the islands of the Furneaux Group) and a large promontory to the west (its tip now King Island). Unlike previous inundations, this flooding of the plain was to have major consequences for both the people and the thylacine.

Abel Tasman was not the only mariner fooled by the mountainous, formidable bulk of the place he named Van Diemen's Land in 1642. The French and English expeditioners and adventurers who called in to its shores over the next century and a half were equally ignorant of the fact that it was an island. It seems inconceivable today that the 200 kilometre wide Bass Strait could have been missed by those sailors. But that was the nature of maritime voyaging, a hazardous European occupation undertaken in the names of science, trade and national ambition.

The establishment of a convict colony at Sydney Cove in 1788 meant that the east coast strip of *Terra Australis Nullius*, together with its mysterious southern part, became the formal possession of Britain, as did Norfolk Island, further east in the Pacific Ocean. Within ten years George Bass and Matthew Flinders had proved that Van Diemen's Land was in fact an island and plans were made to settle it at once. France and England were at war and French captain Nicolas Baudin's scientific expedition to the island was considered to be evidence of Napoleon's designs in the area.

There were other reasons. Seals and whales abounded in the cold waters. Timber was plentiful, for shipping back to England for building ships and houses. And another profound technological development—industrialisation—had produced as a spin-off decidedly uncivilised crime-ridden cities, London, Birmingham and Liverpool in particular. Poverty, opportunism and draconian laws against even the most insignificant theft or misdemeanour resulted in vast numbers of convictions. Excess convicts could no longer be sent to the North American colonies, because of the outbreak of war, while hulks moored in the Thames as floating prisons could only ever be a temporary solution. Hence the settlement at Sydney Cove, followed by the even more satisfactory revelation of the existence of a whole island, ideally suited to become a prison.

In September 1803, a ship bearing 49 people, of whom about half were convicts, hove to in Storm Bay, where the Derwent River meets the Southern Ocean. Under the command of Lieutenant Bowen, they set up camp a few kilometres upriver at a site known as Risdon Cove.[1]

The party soon moved across the river to a more suitable base at the foot of the Table Mountain, subsequently Mount Wellington, where there was plentiful water, good timber and an islet in Sullivan's Cove where ships could be loaded and unloaded. This move was on the orders of Lieutenant-Governor David Collins, now in charge of the fledgling settlement. Collins had relocated from Port Phillip, later renamed Melbourne, which he had declared unsuitable as a settlement.

Accompanying the Collins party was the Reverend Robert Knopwood. Just a few months after Hobart's establishment he bluntly recorded a sighting of a thylacine in his journal account of the plight of some escaped convicts:

> Engaged all the morn upon business, examining the 5 prisoners that went into the bush. They informed me that on the 2 of May when they were in the wood, they see a large tyger; that the dog they had with them went nearly up to it, and when the tyger see the men which were about 100 yards from it, it went away.

Knopwood's journal is the most complete account of the mud and toil of the tiny, convict-built settlement at the southernmost edge of the world. Getting about on a little white pony, he was a 'quaint character, fond of shooting and fishing, especially fond of good living and the society of boon companions'.[2]

The island remained almost wholly unknown, its interior a formidable challenge. By night, fearsome screams came out of the thick woods. The earliest settlers were not to know that those were the calls of the island's other carnivore, the scavenging *Sarcophilus harrisii*, a creature they named to reflect their fear: the devil. And by day, in that first summer of 1803–04, the sky was blanketed with vast plumes of acrid smoke, as the 'natives' practised land management through selective burning, using the fierce heat of the malevolent northerly wind to speed the process.

What must it have been like clinging so precariously to the unknown? Collins wrote this revealing descriptive account, in which nature surely has all the power:

> The extremity of Van Diemen's Land, like that of Tierra del Fuego, presents a rugged and determined front to the icy regions of the south pole . . . It abounds with peaks and ridges, gaps and fissures, which not only disdain the smallest uniformity of figure, but are ever changing shape as the point of view is shifted. Beneath this strange confusion, the western part of this

Robert Knopwood, the first Church of England Chaplain of Van Diemen's Land. Painting by TG Gregson, 1804. (State Library of Tasmania Heritage Collections)

> coast-line observes a regularity equally remarkable as the wild disorder which prevails above. Lofty ridges of mountains, bounded by tremendous cliffs, project from two to four miles into the sea . . .[3]

During their circumnavigation Bass and Flinders had entered the island's northern Tamar River and reported favourably on prospects for a second settlement there. At this time, the Norfolk Island colony was being wound down; its convicts, free settlers and governor, Lieutenant Colonel William Paterson, consequently founded Port Dalrymple; the settlement was later moved down-river and named Launceston. In this way the island was divided into two 'provinces': the quaintly named Cornwall and Buckinghamshire.

Paterson wrote the first extensive description of the mysterious Vandemonian 'tyger'. It appeared in the *Sydney Gazette and*

New South Wales Advertiser of 21 April 1805, which prefaced his description with its own:

> An animal of a truly singular and novel description was killed by dogs the 30th of March on a hill immediately contiguous to the settlement . . . It must be considered of a species perfectly distinct from any of the animal creation hitherto known, and certainly the only powerful and terrific of the carnivorous and voracious tribe yet discovered on any part of New Holland or its adjacent islands.

Paterson's description reads, in part:

> It is very evident this species is destructive, and lives entirely on animal food; as on dissection his stomach was found filled with a quantity of kangaroo, weighing 5lbs, the weight of the whole animal 45 lbs . . . length of the eye, which is remarkably large and black, 1¼ inches . . . from the shoulder to the first stripe, 7 inches; from the first stripe to the extent of the body, 2 feet; length of the tail, 1 foot 8 inches; length of the fore leg, 11 inches; and of the fore foot, 5 inches; the fore foot with 5 blunt claws; height of the animal before, 1 foot 10 inches; stripes across the back 20, on the tail 3; 2 of the stripes extend down each thigh . . . on each side of the mouth are 19 bristles . . . 3 fore teeth in the upper jaw, and 6 in the under; 4 grinders of a side, in the upper and lower jaw; 3 single teeth also in each; 4 tusks, or canine teeth, length of each 1 inch . . . the body short hair and smooth, of a greyish colour, the stripes black . . . The form of the animal is that of the hyaena, at the same time strongly reminding the observer of a low wolf dog. The lips do not appear to conceal the tusks.

Paterson, an avid naturalist and optimist, confidently predicted that within a few years his tiny settlement along the Esk and Tamar Rivers would be thriving. It would do so thanks to cattle and sheep. Far south in Hobart Town (a brave walk of eight days and nights along the midland plain, or two days on horseback), free settlers were hoping for the same as they expanded timidly north along the Derwent, planting potatoes and wheat. In both places starvation became the grimmer reality. Indifferent soils, poor animal husbandry, lack of supplies from Sydney—itself suffering hunger—and aggrieved natives put paid to visions of bounty and splendour. The European presence may have intrigued some bands of Aboriginal people, but for others the invasion, though still minuscule, had an immediate impact. Not only did their hunting grounds become cut off; the very food itself, kangaroo and emu, began to be taken in ever larger quantities by the invader.

The inevitable clash over land and its worth set a precedent for mistrust and mutual suspicion. Needless to say, various levels of relationships between invader and indigene would develop, some decent, but 'the settlers were warned to be on guard against the people of the island who, from what were described cryptically as circumstances that had formerly taken place, had become very much irritated against the Europeans'.[4]

What was the nature of the place they were attempting to tame in order to turn it into a prison? The ironies inherent to the task became very real in the first few years.

Irony one: lack of food became so acute that Hobart Town's convicts had once again to resort to theft in order to survive. Collins's reprisals included flogging and hanging—but these achieved little. Almost everyone was suffering. Lack of seeds and livestock and Aboriginal hostility hindered attempts at expanding the feeble little settlement. So awful were circumstances that the pigs had to be fed on scraps from the whaling vessels which put into the Derwent; the pork began to taste of lamp oil.

Irony two: convict servants were given guns and dogs in order to go into the bush to secure kangaroo and emu, thereby saving the colony. They took to the task with relish. Soon enough they, and many others, were wearing kangaroo shoes and cloaks and in time 'tiger' hats may have become sought-after items.

Irony three: the European invaders were forced to adopt survival techniques not far removed from those of the indigenous people, subsisting and making full use of what the island provided. The Europeans did so, but in achieving this success lost sight of its intrinsic value.

Irony four: a new breed of colonist, the bushranger, soon began to terrorise both native and settler. To the extent that Hobart Town and Port Dalrymple might call themselves genteel—for the benefit of the Crown—this description would always be undermined by commoner observations on the ground. Those two little hot spots of Van Diemen's Land, in their second or third summer, were each a rum economy, 'in the hands of what one newcomer described as a set of the biggest rogues and scoundrels in the world'.[5]

How might such a base society at the top of the food chain treat a shy carnivore itself at the top of a food chain determined over hundreds of thousands of years? The thylacine had no specific meaning for Bowen, Collins, Knopwood and their people. It was merely a creature somewhere out there, as vague as the large interior was unfathomable. The Tasmanian emu may have helped to save the fledgling colonial outpost from starvation, but in doing so its fate was sealed. Easy to catch or shoot, and particularly vulnerable to humans and dogs while sitting on the nest eggs for some eight weeks, it soon became extinct—a tiny sign of 'progress', perhaps, towards controlling the unpredictable island. There were those who warned, to no avail, of the emu's likely extinction, but such attitudes had no place in the young colony.

Today, the known local remains of the Tasmanian emu amount to a few eggs and one feather.

Author James Boyce in his award-winning 2008 history book *Van Diemen's Land* writes of an irony associated with the extinction of the Tasmanian emu:

> Few Tasmanians know there was once a Tasmanian emu. The emu and the forester kangaroo (which also narrowly escaped extinction on the island and is confined to regions far from the capital) on Hobart's coat of arms are seen as quaint decorations, not as representative of the city's birth. The motto beneath the posing animals, *Sic Fortis Hobartia Crevit* ('Thus in strength did Hobart grow')—first used by Robert Knopwood in his diary on the last day of 1804 when the chaplain, with evident pride and gratitude, tallied the number of animals his convict hunters had killed during the first year of settlement—has been reduced to an obscure piece of Latin unrelated to the bounty to which it refers.[6]

6

BEFORE THE FALL: LUTRUWITA

> My Grandfather had a farm there. He and his sons were doing some clearing land, burning stumps at night, and I went with them while they stoked the fires. There was a gully just a way from where the fires was, and these Hyenas or tigers would come and watch, we could see there [sic] eyes shining, we had a clear look at their outline if we went towards them then they would dissapear [sic] but would come back, when we went home they followed us but didn't harm us. Dogs were afraid of them. They done no harm to any stock I can recall.
>
> S. Mitchell, Cosgrove Park

The island upon which the thylacine became isolated had geographic and climatic features rendering it doubly unique. Not only was it part of an ancient floating laboratory but subsequently, through various Ice Age-induced phases as a true island, it developed diverse temperate conditions supporting a variety of life forms. As will be shown, the thylacine population may have been able to benefit significantly from the presence of the Palawa Aboriginal people on the island they knew as lutruwita over the past 40 000 years.[1]

The island is an extension of mainland Australia's Great Dividing Range. Its mountainous western side receives copious rain delivered by the constant winds of the Roaring Forties. Large areas to the east in these mountains' rainshadow are dry for much of the year. Vegetation types and the animals feeding off them are adapted to this west–east, wet–dry pattern.

Prior to European settlement of the island, lutruwita had enjoyed a seemingly benign existence. There is no evidence of tsunamis, earthquakes, volcanoes, dramatic extinctions or other major natural calamities. A number of natural conditions have operated together to the benefit of the thylacine since the island's existence in its geographically modern form. One is the climate—even during its colder periods and tundra-like conditions, prey was in good supply. Another is the island's size. At nearly 70 000 square kilometres, it is large enough to support a variety of micro-climates and the relative abundance of prey in each. Then there was the absence of quadruped carnivores in direct competition with the thylacine. (Earlier, Tasmania's carnivorous megafauna, specifically *Thylacolea carnifex*, may well have been at the top of the food chain, until wiped out through a likely combination of Aboriginal hunting, fire and adverse climatic changes, none of which impacted negatively on the small thylacine.)

Most importantly, Aboriginal people did not fear and therefore did not persecute the thylacine, nor was it a traditional part of their diet, although most other terrestrial creatures were: macropods, the native cat, possums, wombat, echidna, seals, penguins, swans, emu and more.[2]

Pre-1803 hard data on the animal in Tasmania are virtually non-existent, whereas something is known of the pre-contact Aboriginal tribes, derived from archaeological studies and oral sources, and of Aboriginal people in settlement times, in particular from the diaries of George Augustus Robinson, the controversial Aboriginal–settler 'conciliator'.[3] Because the thylacine's distribution across the island was strikingly similar to that of its indigenous people, a look back at the animal in its natural state becomes possible through what is known of the people of lutruwita in their natural state.

There were a number of Aboriginal languages, resulting in various names being given to the thylacine: *lagunta*, *corinna*, *laoonana* and *ka-nunnah* amongst them. Just one traditional thylacine story is known to be recorded. It is part of the 1830s Joseph Cotton collection as told by the 'high priest' Timler, and attributed to Mannalargenna, 'a great Warrior, Sage and overall Chief of a North-East Coast Federation of Tribes'.[4] It is an insight into possible Aboriginal attitudes towards the thylacine. The story is named 'Corinna, the Brave One'.

> Palana, the little star, was the son of Moinee [the great spirit god, ruler of Trowenna (lutruwita)]. As a boy he loved to wander in the bush and had many happy adventures. One day, however, he had a nasty encounter with Tarner, the big boomer kangaroo.
>
> Tarner was huge and powerful, and in a very short time Palana, even though he was the son of the great Moinee, was in dire trouble. The boomer knocked him sprawling and attacked him with his huge heavy hind feet.
>
> Somehow Palana managed to get up, but when he tried to run away Tarner caught him in his arms and quickly throwing him again to the ground, began to stamp the life out of him.
>
> Palana screamed as loudly as he could, 'Help! Help!'
>
> The echoes chased around the bush, rushing from tree to tree, crag to crag.
>
> A nameless hyena pup, enjoying an unequal chase with Lenira, the Bandicoot, heard the cries. He stopped chasing Lenira, who could not believe his good luck, and raced to help.

Fearless, the pup leaped into the fight, ripping and tearing at the big boomer. Tarner picked up the boy, and backing against a rock, squeezed until Palana felt his life almost ebbing. The Great Kangaroo kept the young hyena at bay with his big raking hind feet.

The smart pup quickly dashed up onto the rock and sprang at Tarner, driving his sharp fangs deep into the big animal's throat. Holding the boy with one forearm, Tarner clutched desperately at the brave pup, trying to break the deathhold he had on the kangaroo's throat.

But the little hyena was there to stay.

Body tense and eyes closed, he concentrated all his strength in a mighty effort to close his jaws. Slowly he felt the flesh and sinew give under the pressure of his grip, and suddenly his teeth crashed together with a loud snap.

The big boomer, staggering and trembling violently, crashed to the ground, taking Palana and the pup with him.

There they lay motionless, exhausted and stunned beside their dead enemy.

Some time later a party of blackmen picked up the unconscious pair and carried them back to camp. The pup recovered first. Soon Palana stirred and looked about him.

There he saw Moinee the god, his father.

Walking up to Palana, the god smiled down at him and said, 'You have done well for one so young, my son. You have come through your baptism of danger bravely and unaffected. In a very short period of time you have passed from childhood and now stand on the threshold of young manhood. So be it.'

Straightaway the little boy arose and stood proudly.

He appeared to ignore his father so intent was his gaze on the hyena pup. Moinee read Palana's thoughts and a look of admiration crossed his stern face.

'From today you will make your own decisions,' Moinee said, 'and you will bestow your own rewards.'

But Palana heard not a word.

Walking over to the little hyena, the boy put his arms around the torn and bleeding neck, gently helping the pup as he rose painfully to his feet on tired, wobbly legs.

Looking into his weary yellow eyes, Palana said 'Truly you are the bravest of the brave. Today you fought not as a pup but as the Wurrawana Corinna, the Great Ghost Tiger.'

Kneeling down beside the pup, Palana reached down to where his blood had run into the ashes of the fire, and with his fingers, mixed the blood and the ashes into a thick paste.

Then, with this thick brown paste, Palana described a number of dark stripes across the pup's back from the top of his shoulders to the butt of his rigid tail, saying as he did so, 'From this day forward, all shall know you as Corinna the Tiger.[5]

In contrast, the island's other sizeable carnivore, the scavenging devil, is called by Timler 'Taraba, the Nasty One'. He is disliked by all because, with his coal-black coat, 'he would skulk around on very dark nights, silent as a ghost, attacking the very small, the very young and the very old'. The contrast with the heroic young thylacine is starkly obvious and suggests a potential for the kind of man–animal relationship that had developed elsewhere with *Canis familiaris*, the domestic dog.[6]

Tasmania has nine bio-regions. The island may also be more broadly described as having six distinct geographic regions, in most of which thylacines were well distributed and possibly abundant, subject to the usual interruptions to the food supply such as prolonged drought or severe cold. The regions are: the north from west to east; the mountainous west; the central plateau; the midland plain; the east coast and south-east peninsulas; and the south-west (where thylacines were probably scarce).

Much of the west consists of cool, wet rainforest, the mountains, valleys and river systems of which are unsuitable not only to a pursuit predator but to its prey as well. This was not always the case, however. At the height of the last Ice Age, about 20 000 years ago, rainforest cover was minimal. The Kutikina cave in the heart of the south-west, on the Franklin River, was home to a human population whose hunting grounds were the nearby tundra plains where wallabies browsed. Wallabies are prime thylacine prey, so they too would have flourished there, until gradual climate warming produced rainforest conditions and caused the wallabies to disappear, followed by their predators.

These spectacular western mountain ranges of Ordovician sediments in places soak up annual rainfall of more than 3500 millimetres, and can experience snow at any time, but are equally vulnerable to the fires that are such a feature of the Australian seasonal cycle. Dense stands of beech, sassafras, King Billy pine, pencil pine, Huon pine, celery top pine and others—many of which are original Gondwanan stock, which is why they're also found in New Zealand and South America—produce some of the planet's tallest trees and a sunless floor thickly covered in decaying matter. The forest near Cox Bight in the south-west contains the world's most ancient living clonal organism, the 40 000-year-old *Lamatia tasmanica* plant (King's Holly), the breeding ground of one of Australia's rarest birds, the orange-bellied parrot, and is home to the terrestrial mountain shrimp, which has a 200-million-year lineage. Lichens and mosses are prolific. Impenetrable woody thickets of *Anodopetalum biglandulosum*, commonly known as horizontal, together with leeches, make the area hard going.

Yet the temperate maritime climate enables these forests to support animals like the devil, the carnivorous spotted-tailed quoll and the forest-specific eastern pygmy possum (which, however, is obliged to enter a torpor-like state to survive winter). Echidnas, pademelons (the small rufous wallaby) and possums also inhabit the less rugged areas of the west. All of the latter are prey for the thylacine, which is capable at the least of inhabiting rainforest fringes.

Salt marshes, dunes, wet sedgeland and buttongrass plains make up a generally narrow coastal strip along the entire west coast, but despite being in the face of the wet, cold Roaring Forties, the strip supported both man and thylacine. It is a good example of the animal's ability to adapt to a somewhat uncomfortable microclimate.

The South-West Tribe occupied territory from the natural mid-coast boundary of Macquarie Harbour to the far south-west, including the large offshore Maatsuyker and De Witt Islands, which were visited seasonally for sealing and muttonbirding. This 450 kilometres of coastline supported just four bands, making up a tribe of up to 250 people.[7]

Not surprisingly, archaeological evidence from middens shows that theirs was a largely marine diet of shellfish, crayfish and seals, supplemented by wombat and wallaby from the narrow coastal plains. The comparatively small human population suggests a similar thylacine pattern, that is, thylacine numbers would also have been governed by modest terrestrial prey populations.

Just as the west coast bands adapted by building fully enclosed huts to cope with the weather, so too the western thylacines might have adapted to their harsher surrounds. Paddle notes that western thylacines may have been darker than their more temperate relatives:

> It has been suggested that both size and background colour were associated with thylacine habitat preference, and hence

> particular sub-populations of the species . . . Such suggestions, that rainforest specimens were darker than those living in the drier, more open central plateau or coast, and that a larger body mass was associated with living in colder, more extreme environments, are not unusual for any mammal.[8]

The Aboriginal peoples' huts were built in groups at intervals along the west coast and were permanent bases. Evidence of them exists in the form of 'hut pits': circular depressions dug into the soil. The dwellings were constructed over these pits, from a framework of poles bent into a dome shape and thatched with grass and bark. Groups of huts were also built on the cold central plateau. Although a retiring and elusive animal, the thylacine was known to be inquisitive about humans, even if this was only when they had food about them. But that is the first step in semi-domestication. Hand-reared thylacines apparently made manageable pets. They did not have the antisocial nature of the devil. It is not impossible to imagine a permanently settled village having a nearby, non-threatening thylacine presence, the animals waiting to feed on offal and carcasses. But any such relationship would have ceased once Aboriginal people adopted dogs in large numbers in the early 1800s.

In contrast to their southern coastal neighbours, the bands of the North-West Tribe numbered perhaps as many as 600 people, making them the second most populous tribe. Five of their eight bands were associated with the extreme north-west, which included a swathe inland to the mountains. This was a particularly fertile area. The coastal lands and islands were rich in maritime foods, such as elephant seals and muttonbirds, and seasonal swan and duck eggs from the lagoons, as well as the usual marsupials, especially possums. The people 'moved seasonally up and down the coast, travelling along well-marked footpaths or roads to gain easy access through swampy country covered with dense scrub', and further east they used to travel 'regularly into the high inland country belonging to the North people, particularly to the Surrey and Hampshire Hills region to collect ochre. They travelled through a chain of open plains kept clear by regular firing'.[9]

These deliberately created alterations to the land, for ease of passage, would inevitably suit animals on the move, in the way that, for example, elephant herds bashing clearings through the jungle create pathways and light and thereby openings for other faunal (and floral) species. It was land management of a kind that encouraged, rather than frightened off, animals. Cruel proof of the thylacine's active presence in this fertile corner of the island was to come later, with the establishment of the Van Diemen's Land Company and its disastrous sheep runs on those same Surrey and Hampshire plains. The Company, headquartered at Woolnorth at the north-west tip of the island, was to institute the first thylacine bounty scheme. The Company, so its directors claimed, was being greatly harmed by thylacine predations upon its sheep.

East of the mountain ranges, the island's next distinct geographic region is dominated by the high central plateau, much of which is 600 metres above sea level, rising to 1200 metres. Repeated glacial activity created many thousands of highland tarns as well as larger lakes, including Australia's deepest, Lake St Clair, described as 'incomparable, the loveliest lake in the world'.[10] Rich soils, snowfalls at any time of the year and severe frosts ensure alpine and coniferous forest conditions at higher levels. The plateau is also resplendent with cushion plants, mosses, pandani (*Richea pandanifolia*, the world's largest heath plant, is endemic to Tasmania), poa tussocks, bog, sedgeland, montane grasslands and, at lower elevations, eucalypt-dominated dry sclerophyll vegetation associations.

Devils, quolls, bandicoots and native hens are notable among the species occupying the colder, higher parts. At more moderate levels forests are interspersed with fire-managed plains of silver tussock grass, resulting in wallabies and kangaroos, in very considerable numbers, grazing the grasses, herbs and taller shrubs occurring on these plains and at the forest edges. Later bounty records show that the greatest number of thylacines killed for that purpose were on the central plateau, although that in no way implies that thylacines were most numerous there. (Many snarers operated there, for the valuable fur industry.) Furthermore, the 'distribution of thylacines bore no relation to altitude. They were found throughout the state, and, if anything, favoured the coastal plains and scrub. However, open savannah woodland was used extensively by thylacines and they were not "confined to the mountainous regions" as was so frequently stated in the literature. This was substantiated by all of the old trappers I talked to about thylacines'.[11]

Questions of thylacine abundance and distribution will never be entirely answered while there are significant gaps in knowledge, such as the size of the family unit and the size of the family home range. Estimates of the latter vary considerably, ranging from a few square kilometres up to almost 300 square kilometres. The Tasmanian devil is known to travel up to about sixteen kilometres at one time in search of food, which suggests an ability in its larger relative to cover considerable ground.

The Big River tribe, the people of the central plateau, occupied some 8000 square kilometres of mostly prime territory. Snarers' records for the plateau, collated by Guiler, indicate an average home range of about 80 square kilometres per thylacine family group of, say, six: two adults and four young in various stages of dependency. Was there, then, a general average of about one hundred families, being some 600 thylacines, on the central plateau?

What is more than likely is that the thylacine population and distribution on the central plateau waxed and waned over long periods of time, governed by both climatic and human influence. Evidence comes from archaeological excavations at two plateau rock shelters. The Warragarra shelter, first occupied by people about 10 000 years ago as forests claimed elevated areas previously under glaciers, is thought to have initially served the function of 'an ephemeral hunting camp'.[12] Faunal remains include wallaby, pademelon, bettong, brushtail and ringtail possums: all standard thylacine prey as well.

By about 4000 years ago, however, warmer and moister conditions had turned the area into largely uninhabitable rainforest. The archaeological layers for this period are blank. Warragarra was unoccupied for some thousands of years; it is reasonable to assume that thylacines were also driven from the region.

Warragarra was re-inhabited as climatic conditions similar to today's again took over. As well, the people of the Big River Tribe of the central plateau began the practice of land management through fire, creating more open vegetation cover at the expense of forests in order to increase browsing prey populations. Drier conditions suit firing. That they were dry is proved by the addition to Warragarra of eastern grey kangaroo bones; the island's only kangaroo is restricted to the drier eastern parts.

Slightly further east, the Parmepar Meethaner rock shelter, in a valley which escaped glaciation, was more or less continuously occupied from about 34 000 years ago, subject to forest movement. Wallaby, again, is the main food item. And, again, for the people of the region firing practices became regular, 'producing corridors of more open vegetation and thus facilitating both increased travel and settlement'.[13] This is another instance of creating conditions favourable to thylacines. It should not be considered remarkable that maps of the island showing locations of tribal bands, of important hunting grounds, and of later thylacine bounty claims, are similar. Thylacines concentrated in those areas where their main prey was most abundant and where climatic conditions were suitable.

The phenomenon of land management through firing is most evident in the third geographic region, the midland plain. This is the flat, broad expanse between Launceston and Hobart, blessed with rivers running north and south, and neatly bounded by hills and mountains to either side known as the Eastern and Western Tiers. Aboriginal people managed it by repeated firing practice, possibly removing the original forest cover and so creating the island's biggest kangaroo hunting grounds, at today's Campbell Town and the Norfolk Plains. The midland plain was shared by bands of the Oyster Bay, North Midlands and Ben Lomond Tribes. (All across the island, band and tribal boundaries were strictly observed. Unwarranted transgressions caused bad blood, and friction between neighbouring bands and tribes was not uncommon. Visiting rights and sharing of resources—food, ochre for body decoration, inebriating cider gum sap—were much more important.)

Some of the island's driest areas are to be found on the midland plain. While many of its inhabitants had no direct access to the coast, the inland rivers were a valuable source of food, augmented by a wide variety of vegetable matter and the ubiquitous kangaroo, wallaby and possum. The rising Tiers, with their lightly wooded cover, were excellent territory for thylacines, being close to their prey source while ensuring a good choice of dens, be they under a rock overhang, in a hollow log or a manufactured nest under a scrub canopy.

The eastern coastal fringe rising to the Eastern Tier provided good thylacine hunting territory. Settler landowners on the mid-east coast were later to claim that thylacines were in such plague proportions that, as had been the case for the north-west Van Diemen's Land Company, they were threatening the entire sheep industry. A letter written in 1826 and sent from Oyster Bay stated bluntly that 'tigers are plentifull amongst the rocky mountains and destroy many sheep and lambs'.[14]

Thylacines were also known to favour both the Forestier and Tasman Peninsulas, with their landscapes ranging from long beaches—thylacines liked beaches—to small plains, the lightly covered Tiers and thickly forested slopes. One likes to imagine that thylacines watched the antics of Abel Tasman and his men in December 1642, anchored off the Forestier Peninsula, territory of the Oyster Bay's Pydai-rerme band.

Thylacines would inevitably have been caught in wallaby snares or kangaroo-spear traps: what to do with such a catch? Some Aboriginal bands ate the thylacine, according to Robinson's diary. Others may have respected the animal, as the story of Corinna seems to suggest, placing not only a taboo on eating it but possibly some act of contrition for killing it. The true relationship between the lutruwita peoples and the thylacine will probably never be known, but it features in no midden or rock shelter as faunal remains, and the possibility of its having occupied a special and therefore protected place in the island's human-managed ecosystem is a compelling one.

7

A LAND IN NEED OF TAMING

Several sheep and lambs were dead; others had lumps of flesh hanging from their bodies and their throats oozing with blood. Some were tangled in fences and torn to pieces. In all, forty dead and others had to be destroyed. Surely this was the murderous work of the tiger that had been sighted on the property only a few days before . . . The following morning, very early, there was a commotion down on the flats. Sheep were bleating and running in all directions. It was not yet light enough to see the tiger that was attacking them but two gunshots disturbed the dawn. 'He won't kill any more sheep,' said my father as he approached the culprit. A large shaggy dog. The other one was wounded and got away . . .

OWEN HOUSE, EAST DEVONPORT

A few nights ago a hyena tiger, an animal so rarely seen in this Colony, but of the largest size, was found in the sheep-fold of G. W. Gunning, Esq. J.P. Coal River.—Four Kangaroo dogs, which were thrown in upon him, refused to fight, and he had seized a lamb, when a small terrier of the Scotch breed was put in and instantly seized the animal, and, after a severe fight, to the astonishment of everyone present, the terrier succeeded in killing his adversary.

(*HOBART TOWN GAZETTE* AND *VAN DIEMEN'S LAND ADVERTISER*, SATURDAY 8 AUGUST 1823)

Even after twenty years of settlement in Van Diemen's Land, written references to thylacines were few. The animal, it seems, was rarely sighted by the European newcomers. According to one account, just four were seen in the Hobart region over that period.[1] Yet contradictions were already becoming apparent. Was it rare or plentiful in the wilds? Was there an 'opossum-hyaena' and also a 'panther'? Did either or both of these prey on sheep?

Early in his diary the Reverend Robert Knopwood had written, 'I make no doubt but here are many wild animals which we have not yet seen'; he had also noted that absconded and recaptured convicts were the first to see the thylacine. Given that most of these men were poorly educated, and possibly prone to exaggeration, it is little wonder that a myth might have developed around the wraith-like stripey beast. Nocturnal, huge of gape, capable of menacing hiss-like coughs and barks: how tempting it must have been for these hard men to embellish their tales of the creature.

As the bush's very wildness rendered it friend and ally to increasing numbers, however, the colony's de facto seat of power transferred from Hobart Town and Port Dalrymple to the huts, caves and territories of the bushrangers, and Van Diemen's Land slid into lawlessness even as its permanence as a European settlement became a tangible reality. Gangs of escaped and emancipated convicts and corrupt soldiers, all well armed and attended by packs of hunting dogs, helped themselves to the produce of the strips and patches of river-hugging land cultivated by free settlers. They destroyed, raped and stole almost at will, compounding the difficulties already faced by the more law-abiding. Worse, the gangs were given considerable comfort by some free settlers, so that the job of the authorities was made that much more difficult.[2]

Van Diemen's Land was not a place of sentiment, less still a place where an animal of little economic value might be judged on its merits. Instead, the thylacine came to be branded with a reputation from which it was never to recover. If it is true that dingoes led to its mainland extinction, then it is also true that the spread of feral dogs played a significant part in its downfall in the island. Kangaroo dogs were bred in profusion because of their importance in securing food for the young colony. In the early years in particular, dog theft was a very serious offence. Bushrangers and Aboriginal people kept dogs in large packs. Many went feral. The extent of their harassment of the fledgling sheep industry, for which harassment the thylacine was held at least equally responsible, is yet to be quantified. An 1892 opinion is instructive, written by a mainland correspondent 'Bruni' in the course of reporting on a meeting of the Australasian Sheepbreeders' Association:

> From what I have been able to gather of the nature and habits of the Tasmanian tiger I feel satisfied that one vagrant tame dog will do more mischief in one night in the way of sheep-killing than the tiger will do in a week. Few people are aware of the amount of damage done by tame dogs among sheep. If an accurate return of the losses thus caused in Australia could be obtained, the result would be a great revelation to most people.[3]

The thylacine represented one element of the wildness to be tamed, and its turn would come. Until then, bushrangers, Aboriginal people defending their lands and way of life, crop failures, disease, drought and licentiousness were enough to be going on with, let alone that the struggling island colony continued to receive the dregs of the New South Wales convicts and was in other ways poorly treated by the parent colony. Yet change would inevitably occur and it happened through a combination of the island's natural bounty, its charms from a distance and the introduction of a horrifying system of isolated penal settlements. The grand plan, of ridding England of its convicts and then using their labour to supply the mother country with quality Van Diemen's Land timber, whale products and then vast quantities of wool, meant millions of sheep across the island. Had that grand plan involved potatoes, apples or cattle instead, the thylacine would have been spared its erroneous and fatal 'sheep killer' tag.

Important though crops were (wheat, barley, beans, peas, potatoes) livestock—particularly sheep—played a far greater role in opening up the colony. The first boatload of sheep in 1803 numbered a few dozen. By 1820 there were about 200 000 grazing the midlands between Hobart and Launceston, and by 1830, when the first sheep-protecting thylacine bounty was introduced, the total was over a million—now also spread across the north and north-west. Conditions were particularly favourable along the vital link between the two population centres, the midland plain, so much so that Surveyor-General Evans marvelled that the island appeared to have been in a state of civilisation and cultivation for centuries—which of course it had, thanks to Aboriginal fire management practices.

Evans's 1822 *Description of Van Diemen's Land*, with its subtitle *With important hints to emigrants, and useful information respecting the application for grants of land; together with a list of the most necessary articles for persons to take out* proved a highly effective marketing tool, going into a second edition in 1824 and being translated into French and German. Free settlers, mostly British, began to arrive in numbers which significantly increased a population which, at the time of the book's first publication, was just 1500 souls. Van Diemen's Land was becoming a viable alternative to New South Wales. Indeed, Evans wrote of its 'great superiority' over the parent colony:

> Large tracts of land, perfectly free from timber or underwood, and covered with the most luxuriant herbage, are to be found in all directions . . . These tracts of land are invariably of the very best description, and millions of acres, which are capable of being instantly converted to all the purposes of husbandry, still remain unappropriated. Here the colonist has no expense to incur in clearing his farm: he is not compelled to a great preliminary outlay of capital, before he can expect a considerable return.[4]

Such glowing references were sufficient to encourage the London-based Van Diemen's Land Company shortly thereafter to embark on its disastrous sheep farming enterprise, on vast acreages then totally unsuited to pastoralism. But, as will be seen, thylacines came to be demonised for the Company's failures. It is true that Evans tempered his glowing account with the assertion that 'the panther . . . commits dreadful havoc among the flocks', but the fact is that at the time of that book's publication there had been just two recorded accounts of thylacines attacking sheep.

How economical was Evans being with the truth? Writing not much later, historian John West had a rather different view of that same period:

> The narrow grants and wretched homesteads of the emancipist cotters, the sole farmers at this time of immigration, presented but little to please. The settler, whose imagination pictured the rustic beauties and quiet order of an English farm, saw unfenced fields of grain, deformed with blackened stumps; a low cottage of the meanest structure, surrounded by heaps of wool, bones, and sheepskins; mutton and kangaroo strung on the branches of trees; idle and uncleanly men, of different civil condition, but of one class; and tribes of dogs and natives. No green hedges or

> flowery meadows, or notes of the thrush or nightingale; but yet there was the park-like lands, the brilliant skies, the pure river and the untainted breath of morning.[5]

The 'taming' of the land by the colonist authorities and free settlers went ahead in phases. To begin with there were the bushrangers. The island's third Lieutenant-Governor, Colonel William Sorell, took up office in 1817. Deemed 'a person of firm character who would check the turbulent elements in the colony',[6] he was determined to end outlawry. But such was the impoverished state of his fiefdom that one of his first actions was to do something about all his police officers going barefoot. The

Kangaroo dogs were wolfhound-greyhound crossbreeds, resulting in a size, strength and speed required to bring down bounding kangaroos for food. Unlike 'ordinary' dogs, they were said to be a match for thylacines. Kangaroo dog owned by Mr Dunn of Castlereagh Street, Sydney, *1853, painted by Thomas Tyrwhitt Balcombe.* (Courtesy Mitchell Library, State Library of New South Wales)

government arsenal was also bare, its guns having found their way into the hands of the busy bushrangers. Sorell battled particularly against the infamous bushranger Michael Howe and his gang; not only were an uncomfortable number of free settlers in cahoots with the gang, no less a person than Knopwood was dobbed in as having had dealings with Howe. (Sorell's enquiry into this embarrassing possibility came to nothing when a number of vital documents disappeared.)

Sheep thefts, by both white and black people, with continuing feral dog depredations, were causing infinitely more harm to the struggling industry than thylacines. But the exponential increase in sheep numbers continued and the colony expanded, if slowly and violently, until the bushrangers were no more. The offer of free land grants proved irresistible to many in crowded industrial England, in the wake of the Napoleonic Wars.

Convicts were next. They required taming of a different order. In 1821 remote Macquarie Harbour on the wild west coast was chosen as a penal settlement: out of sight, out of mind. The area contained huge stands of valuable Huon pine and also coal seams for the recalcitrants to work. The administration of Sarah and Settlement Islands, the convict enclaves within the harbour, was unspeakably barbaric. Then in 1824 George Arthur became Lieutenant-Governor. Where Sorell's target had been outlaws, Arthur chose convicts. He would crush their spirits. Independence from New South Wales handed Arthur considerable power and in his mind's eye he saw two Van Diemen's Lands, one a gaol for depraved convicts, the other a genteel settler society made prosperous on the sheep's back. The former had to be absolutely separated from the latter for such a twin society to function properly. He set about achieving that through the establishment of the Port Arthur penal settlement on the south-eastern Tasman Peninsula.

But there was an obstacle of another kind to exercise him as well. The spread of sheep and their pastoralists, and the investigation of most parts of the island by the colonists, had finally convinced the Aboriginal people that their heretofore generally passive resistance had amounted to nothing. Arthur therefore inherited a war, which was to result in near-genocide, a bizarre by-product of which was the voluminous diaries of George Augustus Robinson, chronicling the lives of those being exterminated. These diaries form by far the most complete written source of knowledge of Tasmania's Aboriginal people. Robinson's diaries, as we've seen, also recorded numerous interactions with thylacines on his travels around the island.

Loss of land, deaths from introduced diseases, and theft of their women and children by sealers and settlers, had already

Woorady, *by Thomas Bock, c. 1832-3. A member of the Bruny Island Nuenone band, Woorady married Truganini, who would be the last living full-blood Tasmanian Aboriginal person.* (Collection Tasmanian Museum and Art Gallery)

greatly diminished the Aboriginal population. Even so, for the six years from Lieutenant-Governor Arthur's appointment to 1830 they fought a stiff war of resistance in a vain effort to repel the European invasion—but during this period the settler population rose to some 30 000. A military operation put in place by Arthur, known as the 'Black Line', was intended to round up all remaining Aboriginal people and so bring hostilities to a close. Just two individuals were captured. But that hardly mattered; by then there were fewer than 300 Aboriginal people left in the island.

As part of Arthur's effort, Robinson then undertook his conciliation mission. His writings render him something of a Samuel Pepys of Van Diemen's Land.

These thylacine references are extrapolated from Robinson's diaries and represent by far the most comprehensive early written accounts of the animal:

1830

17 August— . . . This is a dreary country. The soil is covered with a coarse wiry grass and moss, and the shepherds said the sheep could not eat it but for its being wet. In travelling over the Surrey Hills saw several skeletons of sheep; McKay [convict servant] said the excessive cold and wet together with the great number of hyaenas kill them . . .

22 August— . . . At 9 am proceeded on with the whole of the people to Middlesex Plains. The kangaroo bounded before us as we passed. Heavy rain, hazy weather. On leaving Epping Forest came on to a small open plain and as we passed here McKay descried just before us several hyaenas in search of game. We gave them chase and the dog Paddy attacked and killed one, which proved to be the female. She had two fine pups, which made off though closely pursued by McKay. Travelled on and stopped on

the side of the Dove River, whilst McKay and Woorrady[7] went to hunt kangaroo. Platt skinned the hyaena, which was very lean from the feeding of pups . . . With difficulty made a fire and after eating rather hearty of kangaroo was tolerable comfortable. The whole of this country abounds with game . . .

24 August— . . . Hazy weather. Proceeded across some sword grass plains, forded small river and ascended some hills. We commenced our day's travels in tolerable wet clothes. Saw some native huts which was very old; also crossed an old track of the natives. The kangaroo bounded away through the forest as we passed along . . . saw several carcases of sheep which had been killed by hyaenas or wild dogs . . .

25 August— . . . Walked out with Mr Robson [Van Diemen's Land Company] and viewed the bridge over the Emu, which is badly constructed. Saw an excellent black coarse granite—Mr Frankland [Surveyor-General] took a specimen of it when he accompanied His Excellency to this place. Sent a letter to Mr Hellyer and a native basket. Mr R consented for McKay to have 7/- out of the store for a bitch hyaena (5/- being allowed for a dog) . . .

29 September— . . . Travelled along the acclivity of several hills. Purposed crossing over a tier to the coast, but finding it exceedingly bad travelling I proceeded down the Mersey and then along the coast to Port Sorell. From the Mersey to Port Sorell the coast consists of sandy beaches with rocky points, and is frequented by hyaenas . . .

30 October— . . . The whole extent of coast from King Georges Point to St Helens Point is heathy . . . Hundreds of beautiful heaths of variegated hue and all in blossom and sending forth their exhalation, some in scent like the hawthorn, together with the immense number of paroquets and kangaroo, and the small

rivers with the numerous swans swimming stately along, rendered the scene delightfully pleasant. I had now got somewhat accustomed to carrying a knapsack or otherwise the pleasure of the scenery would have been greatly marred. Saw a hyaena and a number of forest kangaroo . . .

1831

22 November— . . . Whilst travelling along the east bank of the Derwent the black boy Lacklay discovered the lair of an hyaena with three fine cubs, which he killed before [I] saw them; the old one got away. I was exceedingly displeased at his killing those animals as I had been desirous to procure some of those animals alive, and those were three fine cubs beautifully marked and a fit size to tame. Umarrah[8] and his wife carried away the carcases of those animals and purposed eating them, which is singular since there was abundance of kangaroo and those animals are carnivorous . . . Had a fine north view of Mount Wellington at Hobarton, the sight of which occasioned an emotion more easily to be conceived than described . . .

1832

14 July— . . . The native men went away to hunt. In my way, on ascending a small hill, observed a bitch hyaena, and a full grown pup run away before me. I and my servant stopped and I stood gazing upon those animals as they fled; their tails stood erect. Whilst looking at them suddenly started up close by where I stood two other full grown cubs who ran about scenting the ground. They performed several revolutions and loped over rocks and seemed in great agitation. They came close to where I stood but did not observe me. At last they got on the scent of the old bitch and run off. It would appear that they were asleep when the bitch run away and on missing her when they awoke commenced

searching about. I proceeded and intercepted one of them on his return back . . .

15 July— . . . I now distributed among them presents of beads, knives, scissors &c, with which they were much delighted. Mounted some of them on horseback and otherwise amused them. My people proposed to hunt, remaining at this place for the night. I however prevailed upon them to move on at once towards Mount Cameron and encamp. To this the strangers agreed and we set out on our way to Wob.ber.rick.er. I told them about seeing the hyaena, at which they laughed and said they had speared plenty . . .

29 August— . . . Proceeded on my journey. Yesterday the natives was in ecstasies of joy; on enquiry they told me warm weather was coming and showed me the trees in blossom which had been the cause of their excitement . . . Saw an hyaena at Mount Cameron which had been recently killed. Suppose it to have been killed by the natives who it is more than probable had been following on my tracks since my departure. The New Hollanders skinned the hyaena for the purpose of carrying it to Cape Grim to get the ten shillings reward from the Company. Pleasant weather throughout. New moon . . .

1833

30 August— . . . There was plenty of ducks and crows. Rain at times. The rivers had shifted. Caught kangaroo and killed one hyaena on the sandy beach. The hyaena is called mannal-argenna (east coast), cabberrone-nener, by the Cape Grim, lowenin, by Jenny (north coast) clinner, by the Cape Portland warternoonner, by the Brune cannenner, and by the Oyster Bay larnter . . .

1834

25 January— . . . Say nine or ten dogs and hyaena together had been caught in a month . . .

6 April— . . . *Hyaena hunt:* Just before I got on to the long sandy beach and whilst we were walking along the top of a hill, my dog Fly scented game and run to the bottom. I watched her and in an angle of some rocks with a sandy beach saw a large hyaena at bay with the dog. My son at the same time run down to the dog's assistance and saw three young cubs who with the bitch had been feeding on the carcase of a kangaroo. The unexpected visitant surprised the animal, when she run up the hill followed by the dog. We shouted and the animal alarmed run over the hill, thence across a plain and along the acclivity of another hill for a considerable distance, followed by the dog who from the commencement of the chase had kept close to the animal occasionally biting its rump, when the hyaena would turn around and give chase to the dog and anon pursue its way, and at the same instant she turned the dog was biting her. Another large dog with us and which belonged to the old chief, would not approach it but barked. Black Dick and a white man came up and I sent them down to the dog's assistance (for this was only the second one she had seen and very large). When the men went to it it run at them. At last it fell when the dog Fly sprung and fastened on its throat, and the two men killed it and brought it to me. It was the finest hunt I had seen and a good run over clear ground. My son and some others of the men was hunting the cubs; my son caught one alive and two escaped after being chased. I brought away the skin. This beast was surprised in feeding on game; it was not her lair. I imagine she was making for her lair when chased by the dog. Perhaps the dog hyaena was at the lair, or hunting more game. It might have killed this and brought the bitch to it.

It was the finest I had ever seen, with a head in the shape of a fox. (Describe this hunt hereafter minutely; it will be interesting.) . . .

4 June— . . . a severe frost this night, the first I had seen this winter and the ice a quarter of an inch thick . . . The [Van Dieman's Land] Company have removed their flocks; they have had no increase but declension of stock. Mr Chitty said the Company could not keep sheep here, that if they were to fold them they would soon die, that it was only moving about that kept them alive. Wild dogs and hyaenas are numerous. Watchmen were kept to look after them . . .

18 June— . . . Showery weather. The natives searching for indications of the aborigines . . . Jack said he saw the hyaena hunting a kangaroo on the scent like a dog. He ran and speared him in the tail, and the dog caught him by the neck. They brought him home. It was a bitch hyaena. Had it skinned and the skull saved. Busy since my sojourn at this camp in writing journal and in conversation with the aborigines. God grant I may soon meet the aborigines . . .

21 June— . . . The cause of this bad weather is attributed to the circumstance of the carcase of the hyaena being left exposed on the ground and the natives wondered I had not told the white men to have made a hut to cover the bones, which they do themselves, make a little house . . .

23 June— . . . It was not my wish to encamp at this short stage but the natives were tiresome and murmuring and wanted to stop. Penderoin was walking in advance of me when he discovered a tiger cat asleep in its lair, which was under a tussock of grass and made of old grass torn in pieces. The youth run his spear through it, it was a female hyaena . . .[9]

Robinson has long been condemned for his role in the Tasmanian Aboriginal genocide. Yet his diaries show empathy with both people and their land. His story is part of the island's peculiar tapestry woven of beauty, abomination, destruction, regret and much else besides. Its paradoxical nature could almost be summed up in attitudes to the thylacine—as, for example, witnessed a decade later by Louisa Anne Meredith, whose astute writing and paintings allowed intellectual Britain a glimpse of a different Van Diemen's Land. One day in 1846 a shepherd presented her with a juvenile thylacine. She wrote:

> He had the animal secured by a chain and collar, and when it was to be carried off, slipped a strong bag over its head and shoulders, pushed the hind legs in and fastened it. I pitied the unhappy beast most heartily, and would fain have begged more gentle use for him; but I was compelled to acknowledge some coercion necessary, as when I softly stroked his back (after taking the precaution of engaging his great teeth in the discussion of a piece of meat) I was in danger of having my hand snapped off.[10]

Thomas Bather 'T.B.' Moore, a legendary Tasmanian track cutter, prospector and scientifically minded explorer, occupies a small but unusual place in the thylacine story, being the maker of one of the few known thylacine artefacts. And he is relevant to this chapter because of his tireless and gruelling solo treks through dense wilderness, cutting tracks to further open up the rugged west of the island. He also spent many months exploring the remote south-west. Like G.A. Robinson, he kept diaries, which form the basis of Tasmanian author Patsy Crawford's excellent 2020 biography *The Track Cutter: The Life of Thomas Bather Moore*.

Some three decades after Louisa Anne Meredith's encounter with a thylacine on the east coast, Moore was hard at work on the rugged north-west, in the Pieman River area. As described by Patsy Crawford:

> The shoreline and hinterland above and below the Pieman would have been a natural setting and meeting place for the Aboriginal people who roamed that part of the island. As many as ten clans regarded territory along the coast as their home ground . . . The Aboriginal people referred to the Pieman as Corinna, which is generally interpreted as 'young thylacine'. The riverine environment would not have been as natural to the thylacine for hunting purposes as would the plains of the midlands and north-east, but the tiger was known to inhabit the more difficult terrain of the west and south-west and was certainly recorded by Moore in his diary[:] 'Made on to Clearing Camp where tent was left . . . dogs caught a tigre about thirty yards in front of me just after I started & had it killed before I got up to them. They eat everything but the skin which I took time to make a cap of in remembrance of their first striped gentleman.'

The incident so impressed itself on Moore he was moved to write a poem which in part reads:

> Upon the turf there lay quite dead, A beast with stripes upon his coat
> Young Spero bit about its head, While Spiro grasped it by the throat
> So there a noble tigre died, Just as the sun set golden rays
> Shed light upon the mountain side, And victors of that savage fray.[11]

The fate of T. B. Moore's tiger cap is unknown.

Photograph of T. B. Moore wearing his thylacine skin cap, with his dogs Spiro and Spero. (Margaret Elliston collection)

8

TALL TALES, TIGER MEN AND BOUNTIES

Sheep owners in that area never seemed to complain about losing sheep or finding any dead. All the hyinas [sic] I have seen in the wild seemed to be about half grown and about the size of [a] full grown fox terrier dog, fawn in colour, with dark stripes on their sides and back, with a very long tail continued on from their bodies, their tails were half up and down like a tired dog.

K. M. CRAWFORD, ULVERSTONE

The people were fanatically loyal to the Crown and Britain . . . Hedge-rows were smiled upon as a reminder of a Merrie England few if any of them had really ever known, the destruction of the native flora and fauna and its replacement by exotics went on apace, a fact which added to the picture of colonists being merely transplanted Britons.[1]

To the extent that the above observation is accurate, a grasp of the collective psyche of the growing colony becomes necessary in trying to determine the logic behind dislike of the thylacine. It was a consequence of the general attitude that saw the countryside as alien, finding expression in fear stories (the thylacine as a child-snatching vampire); in heroic man-and-dog versus beast stories; and, most negatively of all, in classifying the thylacine as vermin, to be exterminated through the inducement of financial reward, that is, bounties.

The first bounty came about as a result of the ill-fated scheme hatched in distant London. While official survey charts and maps of the mid-1820s have the entire north-west marked 'This Part of the County Unknown', an earlier report on the prospects of settlement in those parts had stated, 'That quantity of good land which was expected about the North West part of the island, is not to be found'.[2] This was hardly surprising, given the generally narrow fertile coastal strip backing on to an interior of dense forests, valleys and peaks. Yet a small group of wealthy entrepreneurs in London convinced themselves, Colonial Secretary Lord Bathurst and Lieutenant-Governor Arthur that a vast sheep-based enterprise should be established there.

In this way the Van Diemen's Land Company came into existence in 1826, with a land grant that was soon to total no less than 350 000 acres (141 750 hectares). It was allotted in great chunks at the north-west tip, named as Woolnorth, and the central north-west, named as the Hampshire Hills, Surrey Hills and Middlesex Plains. The London directors envisaged handsome profits from the fine wool of a quarter of a million merinos, with no further English reliance on the expensive German and Spanish wool industries.

Instead they, and many others, including the thylacine, were to pay for a monumental misjudgement. The climate was too cold and wet for fine wool sheep; sub-alpine grasses were unsuitable fodder; much of the land was uncleared and therefore unproductive; fencing, once built, proved inadequate, as did the inexperienced shepherds, many of whom were convicts; indentured servants disliked the primitive, remote living conditions; sealers, 'vagabonds' and Aboriginal people helped themselves to sheep, as did wild dogs, devils, wedge-tailed eagles and thylacines.

Records show that many thousands of sheep died from cold and starvation during the first few winters, while records of thylacine predation are minimal. Still, in 1830 the resident Chief Agent of the Company, Edward Curr, issued this bounty scheme:

> The superintendant of the Hampshire and Surrey Hills Establishments is authorised to give the following rewards for the destruction of noxious animals in those districts:
> For every Male Hyena 5/-
> For every Female with or without young 7/-
>
> Half the above prices for Male and Female Devils and Wild Dogs. When 20 hyenas have been destroyed the reward for the next 20 will be increased to 6/- and 8/- respectively and afterward an additional 1/- per head will be made after every seven killed until the reward makes 10/- for every male and 12/- for every female.[3]

The clear implication was that the thylacine was the major animal culprit in stock loss. The first bounty scheme operated for only a few years and its kill rate is unknown, but was probably modest. The few records that survive give no indication otherwise. Then in 1836 at Woolnorth a 'tiger man' was appointed to trap and kill thylacines and other animals deemed a threat to the livestock. Tiger men were to be a feature of Woolnorth into the twentieth century. The title gave status, although arguably they were but shepherds with an additional brief to eliminate 'vermin'. They were paid per animal. As well, thylacine and other pelts fetched decent money for the trapper, particularly in winter when they were thicker. Catching was almost always by snaring.[4]

A good account of winter snaring, as a longstanding form of bush employment, is recorded in a fourteen-page monograph published in 2011: *The building and rebuilding of Basil Steers #1 Hut: February Plains* by Margaret Howe for Tasmania's Mountain Huts Preservation Society. Basil Steers was the island's legendary 'last trapper', working the high country. Howe writes:

> In 1942, at the age of 15, Basil followed his father into the mountains for his first experience of winter snaring. Basil would be away for weeks at a time during the winter months of May, June and July, returning home in August. There was a definite art to setting the traps and with different snares used for different animals. Wallabies and pademelons were caught using a springer neck snare, while the larger kangaroos were caught with a wire drag. Possums were caught by a wire noose on a pole.[5]

A second Woolnorth bounty scheme began in 1840—but again, the records give no real indication that either the bounties or the tiger men were necessary at all, and the suspicion must arise that invented or greatly exaggerated 'ravages' of 'savage' thylacines against the flocks conveniently masked the true nature of the Company's pastoral failings.

The second bounty scheme coincided with a major economic slump in the island, which was exacerbated when transportation to New South Wales ceased and all British felons were now sent to Van Diemen's Land, adding greatly to the cost of running the colony. The pastoral struggles at Woolnorth continued, although the company's losses began to be offset as pioneers purchased parcels of its land in this hard country and began a subsistence existence on smallholdings. As the north-west was thus slowly opened up, thylacine territory, already much diminished elsewhere, was further squeezed.

It may be that this growing human presence pushed inland thylacine populations towards and into the 170 000 acres

(68 850 hectares) of Woolnorth and nearby Circular Head. But during the period 1839 to 1850, stock losses recorded at Woolnorth firmly point the blame elsewhere:

Sheep killed by:	
thylacines	40
dogs	660
dogs or thylacines or 'vermin'	1050[6]

The next bounty—which had nothing to do with the Van Diemen's Land Company—lay some 40 years ahead, but conditions for its fatal operations were fast being laid:

> By now the gentry had succeeded in forming a coherent class which appeared to place its impress on the society in a permanent manner. The triumph of the wool industry strengthened their confidence and increased their opulence, and there was constructed in the wool country which stretched between Hobart Town and Launceston, with spurs to the east and west, a number of colonial mansions . . .[7]

This gentry's power and influence increased tenfold when, in the space of two years, transportation ceased, the island was renamed to commemorate the great severing of the penal umbilical cord and, in 1855, newborn Tasmania was constitutionally granted self-government, by the formation of a parliament with a House of Assembly and a powerful, propertied Legislative Council; another nail in the thylacine's coffin.

While Tasmania began to be touted as a wondrous natural sanitarium for worn-out expatriates from such British colonies as India, for the ordinary folk of the island, far removed in all ways from the hubris of the new government in Hobart, life went on, generally as a struggle. An atrocious lack of urban sanitation, pitifully bad education and ongoing lawlessness were just a few

A mid-nineteenth-century German engraving. Few, if any, predators would attack a flock of sheep in broad daylight in the presence of its shepherd. But anti-thylacine misinformation and hysteria obviously travelled well. (Gerard Willems, Hobart)

hazards of colonial life. But these were considered to be almost natural. By contrast, to the collective mentality nature often seemed unnatural—never mind the paradox of that—as exemplified by the tiger-hyaena-wolf-marsupial thylacine. Despite or perhaps because of its being rarely seen by common folk, the animal was feared. Killing a thylacine became something of a rural duty, and also implied considerable bravery.

The following feature appeared in the *Launceston Examiner* of 22 November 1862, headed 'Tasmanian Tigers' and written by J. S. of Forth:

> It [sic] was emerging on to a small plain which is situate about a mile north from the east end of the Black Bluff Mountain, when I saw the bushes in motion a short distance in front of me. I thought the movement was caused by a number of the wombat

tribe, the flesh of which is considered to be good eating by travellers in the bush, to whom meat is a luxury, and I sent forward my young though nearly full-grown dog—a setter spaniel—thinking that he might chase the supposed wombat into a hollow tree, where it could be easily captured. But just at the moment two tigers emerged into the open space, and one of them made off, while the other turned fiercely upon the dog, which running back regarded me with an expression which seemed to ask as clearly as instinct could, whether he was to attack in earnest. Again I sent him forward, and this time he advanced with such a display of determination, that the tiger, after a faint show of resistance, began to retreat, when the dog, seizing it by the tail, suffered himself to be led along at a quickish pace, while I followed with the view of overtaking them, till I stumbled and fell, and narrowly escaped being cut by a tomahawk which I held in my hand. Here the dog rushed back as if to see what was the matter; but I again sent him forward, and the tiger, now finding escape impossible, turned upon its assailant, and both fought with the utmost ferocity for about two minutes, when the dog seizing the tiger by the throat and keeping his hold overpowered it in a few seconds, and I then knocked it on the head, thinking I was doing right in destroying one of a kind of animal which is often very destructive to sheep and lambs. The tiger was sixteen inches high and two feet six inches long from the nose to the tail which was fourteen inches in length.

A short time after the incident above-mentioned I was on the north side of the River Lea, and seeing my dog suddenly bound off, I looked in the direction he had taken, and saw him approaching the largest tiger I had ever seen, and which seemed disposed to treat him with the utmost defiance, rushing at and driving him off whenever he approached within a few feet, and then steadily resuming its course. Thinking, from the size of the tiger, that it might prove dangerous even to a man, should it meet one when without a weapon wherewith to defend himself; and thinking that if it effectually intimidated the dog it might return in the night with its mate—for they often go in pairs—and cause me some annoyance, I resolved upon attempting its destruction, and I ran forward for this purpose, tomahawk in hand; but the moment it saw me advancing it reversed its course, and made for a scrub at the margin of the river; but the dog being encouraged by my voice and presence rushed at and fastened upon it repeatedly, but was as often repulsed with a sharp bite; he, however, so retarded its progress that I was enabled to overtake it when, thinking to despatch it at a single blow, I struck it on the head with my tomahawk and fractured its skull; but, notwithstanding this, it reeling and staggering, fought the dog for fully two minutes, inflicting bite after bite in his neck in rapid succession; and placing its fore feet against the upper part of his breast pushed him off by main strength whenever he fastened upon its throat. While this was going on I could not strike a second blow for fear of wounding the dog, at such a rate did they roll about; at length, however, seeing that the tiger was becoming weak, I seized it by the hind legs, and, placing my foot on the dog to keep him steady, I struck it again with my tomahawk severing the spine of its neck and causing its instant death. The height of this tiger to the lowest part of the top of the back was twenty-one inches; the length of the head and body, three feet four inches; of the tail, seventeen inches. It was a female and had in its false belly four young ones, each adhering firmly to a teat and well covered with hair, and just able to walk. I would have kept these young tigers as specimens of animated nature if I had not found that they would not live on the rough food which alone I had to give

> them. From what I know of the Tasmanian tiger, I believe that whenever it finds itself menaced by superior force it retreats for its den, and if it succeeds in this it faces about, and being secure from attack except in front, it defends itself with the utmost ferocity, inflicting severe wounds on any assailant that ventures to close with it. I have only heard of one contest between a sheep dog and a large tiger, and in that the former prevailed, though not without being fearfully lacerated by the teeth of the tiger. With a kangaroo dog I do not think that any tiger would be able to contend long owing chiefly to the superior mode of fighting of the former. I have never known a tiger to attack a man—although I once knew one to walk up to two in the bush, but in this case the intruder was knocked on the head, before he had shown any aggressive intentions. I have also known a tiger to follow a man for a considerable distance at night, though not attempting to molest him.

Embellished or not, the clear inference is that thylacines were better off dead. The size of this piece, relative to the newspaper's news and features pages, indicates that it was of importance as a Tasmanian story, and the tone clearly shows that it was written for an uncritical, even admiring audience.

But that is not to say that all Tasmanians were of like mind on the subject (just as many native-born settlers did not hark after Merrie England). The Reverend John West, historian and a leading Launceston citizen until he moved to Sydney, might not have read J. S. of Forth's tiger encounters, but would certainly not have been one of his admirers. Ten years before that piece, in his 1852 history of the island, West included this piece, written by Ronald Gunn, in his 'Zoology' section:

> The Thylacine kills sheep, but usually confines its attack to one at a time, and is therefore by no means so destructive to a flock as

> the domestic dog become wild, or as the Dingo of Australia, which both commit vast havoc in a single night. High rewards have always, however, been given by sheepowners for their destruction; and, as every available spot of land is now occupied, it is probable that in a very few years this animal, so highly interesting to the zoologist, will become extinct; it is now extremely rare, even in the wildest and least frequented parts of the island.[8]

The English naturalist John Gould, whose mid-nineteenth century descriptions and illustrations of Australian mammals and birds are unsurpassed, also foresaw the extinction of the thylacine:

> The destruction it deals around has, as a matter of course, called forth the enmity of the settler, and hence in all cultivated districts the animal is nearly extirpated; on the other hand, so much of Tasmania still remains in a state of nature, and so much of its forest land is yet uncleared, that an abundance of covert still remains in which the animal is secure from the attacks of man; many years must therefore elapse before it can become entirely extinct.[9]

Gould's prediction applied to the last frontier, the west, which had remained untouched since the closure of the Macquarie Harbour penal settlement in 1831. The sheer difficulty of getting through the mountains, valleys and dense forests was rewarded when metals, especially tin and gold, were discovered. Surveying and exploration soon covered most accessible areas as gold rush fever took hold. In the Mount Heemskirk area alone (that peak named after one of Tasman's ships in 1642), some 300 gold mining companies had formed by 1882.[10] In that same year, the *Hobart Mercury* of 19 September ran an even more lavish tiger feature, written by 'Oscar', who had been lured west in search of metal wealth:

THE NATIVE TIGER—Having finished a prospecting tour in the southern part of the West Coast of Tasmania, I returned northwards, following the rugged coastline that has so great attraction from a geological point of view. I also experienced that this out-of-the-way place has other interest besides that of inanimate Nature, even in the absence of all human beings, of those dusky forms once belonging to the soil, whom the pale-faced lords of creation have long ago used their superior civilisation to blot out of existence . . . By degrees the evening would become hushed to an almost oppressive stillness, the distant surf, with its monotonous murmur, making the night seem only the more solemn; while resting in the soft ferns with my companions by my side, who are already in the land of dreams, my dog sleeping in a bush of cut grass a short distance from the tent, the fire had burnt down so as to make it only just visible in the darkness. Some undefined noise keeps me wakeful, with my sense of hearing strained to the utmost to make out the meaning of a peculiar creaking sound. At last I felt certain that I heard the breaking of a dried stick coming into contact with a moving body. Gently lifting up the side of the tent, and thrusting out my head and arm, in the low position, as I am lying on the ferns, I can soon make out in the pitchy darkness the appearance of two phosphorus-like orbs, which slowly approach. In a few more seconds I can dimly discern by a light shooting up from a few leaves on the almost expiring fire, the long round body of the native wolf or tiger. I get a tighter grip on the handle of my tomahawk, ready to give a warm reception to my night visitor. I remain motionless, [a]waiting the monster sneaking towards me, expecting every moment that it will take a leap and commence battle; but this seems not to be their general tactic in the night. By inches it is coming nearer; now, surely, it is within reach of my arm. Just then it seemed to perceive me lying under the tent, and gave vent to a low angry snarl. I raised my weapon, and with a swift blow down, as I thought, on the brute's head. But the darkness misled me, and the blow fell short of doing any execution, but near enough to draw a few drops of blood, and thus to stop the brute's further progress. My companions were awakened by the disturbance, and gave vent to some rather unpolite expressions. The dog got the scent of his great enemy and set up a furious barking, and with this salute from bipeds and canine, the feline gentleman made himself scarce, and, unluckily, we had no firearms to persuade him to stop and make a nearer acquaintance. However, the dog rushed after him into the scrub, but soon came back, apparently he did not see the fun of a night attack. Up to the time that this adventure occurred, I was under the common impression that the native tiger would not kill its own food, but was a mere cowardly scavenger. But now I think that this idea can hardly be true, as his aggressive movement in this instance was too decided to be mistaken. Afterwards, in the course of many weeks, we had ample opportunities to learn the habits of the native tiger in this respect, and we observed that it hunts animals with the pertinacity of a pack of wolves on the steppes of frozen Russia. This native tiger is not swift, and is very awkward in turning, but it follows the trail by its never-erring scent, and in the long run is sure of its prey. While travelling one day over some rocky points we quite easily caught a kangaroo, well-nigh run to the ground by the tiger. After despatching the hunted victim we had a sharp look-out for the enemy. In a little while he came along with his nose following the trail. When within some hundred yards he took the alarm, turned tail and escaped into the jungle. My dog, a courageous and very strongly built animal, followed the tiger into the dense scrub. Slowly I made my way through the obstructions, and at last heard the baying signal of 'Wrangel' a long distance off, too far for me to penetrate during the short time the battle lasted. In half an hour's time he came to me bleeding, and with his tail down, showing in his dejected appearance, as plainly as words could tell, about the lost engagement with the enemy. A few kind words had the effect of restoring his wounded feelings, and a good feed from the meat of the killed kangaroo made him all right again. But this constant disappointment in coming to close quarters made me even more determined to be the owner of a striped skin if a chance offered, before I left the coast. I had noticed that the tigers followed our trail, and often when travelling, I would leave my knapsack, and turn back a mile or more, with the object of surprising the animal on the open ground. But he was too wary, and always made for the dense forest. My dog seemed cowed, and never made an attempt to attack by himself. On one beautiful afternoon, I was travelling towards our camp from a trip made that morning, for carrying stores to a newly formed depot. I was returning without any weapon, except my large sheath knife. The road led along the beach, the dog trotting peacefully by my side, or making a playful dart at the pretty little sand larks. But why does he stop the play so suddenly, and pick out one of the many tracks that are so plainly visible on the soft surface of the sand? Impressions of footprints are numerous of all the quadrupeds found on the West Coast. Those of native tigers, Tasmanian devils, and kangaroos are seen everywhere, and many native cats, etc. With a zeal I had not lately noticed, 'Wrangel' giving out a deep bloodhound note, was off and on the warpath. I followed as fast as possible, leaving the sandy beach behind, and turning round a rocky point, there was a glimpse of the dog in full pursuit of some large animal, which took to the scrub. Directly the signal from 'Wrangel' sounded at bay. I rejoiced to know for certain that his enemy had made a stand. The noise came from behind some close, horizontal bushes. Taking in at a glance the position of the surroundings, one could see that no other means existed to come on to the battlefield, except by crawling on hands and knees, under the interlaced bushes. The angry growling of the tiger decided me to push to the assistance of my gallant 'Wrangel' at once, and thus dragging my body over the ground, and under the bushes, I came pretty near running my head against the tiger, who was now placed between the dog and me. No attack had as yet been made from either side. I tried to rise to my feet, but could not, on account of the close branches above; the tiger made an attempt to fasten on to me, with his ugly row of teeth. The dog's courage did not fail me at this critical moment. He made a snap, and fastened on to his opponent's chest, for one instant only; but that served my purpose, to get a chance to make a stab with the bowie knife, and thus I succeeded in cutting the big brute's jugular vein, and his life blood streaming out settled the matter in a few seconds. Nevertheless he had time to make a desperate effort to yaw me as I was quite close up to him. However, I took a short stick in my left hand, and thrust it into his mouth. He chewed it up like pulp. In the meantime I came out from the bushes, and once more stood on my feet, thankful that the encounter had come off so well, and making a promise to be prepared with a better weapon in the next engagement with a native tiger. In a few minutes the body was dragged out on to the beach, and I found it to be that of an old male of very large size, measuring 8ft. in length, including the tail, and weighing about 1cwt [= c. 50 kilograms, a sure overestimate]. I took off the skin, which shows 16 stripes. A young full grown animal rarely shows more than seven or eight. As I had brought no firearms into the bush,

> I did not succeed in 'bagging' any more tigers. In fact, I did not care to venture again so close after them. But I often saw them, though never had the temptation to risk another close encounter. The dog, ever after the first night attack, had been very watchful, and would with his furious barking often wake me up, and on one occasion I saw, on a moonlight night, a big tiger making a hasty retreat from close to the tent, and then take to the water in a broad and deep river, and disappear on the opposite shore. But we never had another actual battle with one of these brutes.

Such heroics, so publicly displayed, must have been appreciated by those sheep farmers who now fixed their sights on the thylacine. The island's urban and rural divide was becoming increasingly marked—in the former through improved health, education and working hours, and amenities such as a public library system, museum, botanical gardens, tramway—while in the rural sector wool and other primary production exports developed steadily. It is tempting to consider that in the boom period of the 1880s, at least urban attitudes towards the thylacine might have been changing for the better, from fear and loathing to indifference to interest. After all, it had long since ceased being the terrible Vandemonian panther, being now displayed in zoos as far away as London. Conditions were favourable for public attitudes to become more benign. Alas, powerful rural interests ensured that would not happen.

Despite the leap in wool production, achieved largely through better control of intestinal parasites, the wool industry remained troubled. Disease, drought, competition from rabbits, predation and duffing (stealing) contributed to significant stock losses. Some averred that stealing was the biggest problem. Poor farm management was also a feature. Flocks properly tended and enclosed in paddocks were far better off than those left to themselves in scrubby, awkward, unsighted hill runs. Losses and the causes of those losses could generally only be estimated, yet despite a clear lack of evidence or proof that the thylacine was a major culprit, in the early 1880s groups of sheep owners in the midlands and in the east began to form anti-thylacine stock protection associations.

Historian James Fenton, writing at the time, made no mention of the thylacine but stated that:

> The 'rabbit plague' (as it is termed) has of late years become a formidable evil on the pasture lands of the island. The midland districts are overrun by this little pest in countless thousands. The sheep and cattle bearing capacity of the runs has consequently been greatly reduced . . . Little short of £100,000 per annum is lost in wool from this cause alone. [Even so,] by the exercise of patient care and skill on the part of the breeders, Tasmanian sheep have been brought to remarkable perfection, and have been eagerly sought after by Australian graziers for the improvement of their flocks.[11]

Paddle notes that 'at no stage in any of the official annual reports on the state of the [sheep] industry during these crucial years was the thylacine mentioned even as a minor problem of significance to sheep farmers'.[12]

The best-known anti-thylacine group, formed in 1884, was the Buckland and Spring Bay Tiger and Eagle Extermination Society, on the south-east coast. Members of the society paid a scale of rates (according to number of sheep owned) and from this kitty £5 would be paid for each adult thylacine and about half that for each cub. Records of its results indicate very poor kill rates of both thylacines and eagles, even though the Society sent a number of specimens to the Tasmanian Museum.[13] Politically, however, this was a powerful rural body able to lobby its House of Assembly member, James Gray, to act on its behalf. The move to render the thylacine extinct was in this way formalised by making it the subject of parliamentary debate.

It was in the 1880s that a range of attitudes towards the thylacine first became evident. Newspaper editorials, letters to newspapers, the concerned voices of naturalists and the staff of Hobart's Royal Society Museum, who were convinced of the animal's growing rarity, indicated that not everyone considered it expendable vermin. Even some of those who did were beginning to appreciate the science of understanding the natural world:

> The Curator of the Royal Society's Museum has just received, through the medium of Mr C. E. Davies, a large male specimen of the Tasmanian tiger, captured by Mr Joseph R. Green, of Nugent. In the letter which accompanied the donation, Mr. Green stated that the tiger was killed on Mr. Joseph Hodgson's estate, between Sorell and Spring Bay, and that its discovery removed the belief existing in the minds of many persons that no native tigers are now to be found in that locality. It is pleasing to note that country residents are beginning to take an interest in the Museum . . .[14]

Unfortunately that report no doubt came in handy for the pro-bounty lobbyists anxious to disprove the reality that thylacines were by now extremely scarce in the area. Other reports directly aided their cause:

> A Tiger capture.—A correspondent writes:—'On the 10th inst., whilst the brothers French, sheepfarmers residing at the Bluff River, in the neighbourhood of Stonehenge, were out on

> their run looking after and setting snares for catching tigers, they came upon one, and succeeded in rushing it into a snare set close by and caught him, and without a single scratch or hurt of any kind have now got the sheep killer safely caged at their homestead, and certainly it is a very nice specimen of the Tasmanian tiger, and is nearly full grown. Already a gentleman has made an offer of £6 for it, but which has been refused by the Messrs. French. These settlers have since they came to reside at the Bluff River killed and destroyed a great number of tigers, and still keep on catching and killing every year, and consequently their good work must be a great benefit to the sheep farmers of the district of Spring Bay, especially to those in the neighbourhood. The loss of sheep during the past 18 months has, however, been great, as the books of the sheep owners can tell, and no doubt is caused by the ravages made by the tigers. Settlers in the district are anxious to know something about the £500 [for the destruction of thylacines] their member moved for at the last session of Parliament, and hope that he will keep the thing alive at the coming session, and thus aid sheep farmers to eradicate the pest from amongst their flocks.'[15]

All the while the private Van Diemen's Land Company bounty scheme and its tiger men had continued to operate, between 1874 and 1887 making a total of 70 bounty payments for thylacine kills.[16] Although more were undoubtedly killed but not presented for bounty, the annual average of five kills suggests low predation. That in itself ought to have cast some doubt on the rural lobby's assertions in other parts of the island that thylacines were almost destroying the sheep industry. But if a bounty scheme were to be extended to the whole colony, backed by government money, by how much would that annual sanctioned kill rise? And for how long would the thylacine population be able to sustain such a rise?

SMITH AND CARTLEDGE

Remind their Friends and the Travelling Public that they

HAVE OPENED A

COMMODIOUS HOTEL

AT

WHALE'S HEAD,

The Port of the Balfour Copper Field, where Visitors will find nothing but the Best. Tourists in search of fun can be accommodated with Kangaroo Hunting, Tiger Shooting, Fishing, Yachting, &c., &c. Horses provided when necessary

COMMUNICATIONS PROMPTLY ATTENDED TO.

This advertisement first appeared in the North West Chronicle *between 1909 and 1912. Whale's Head is now Temma, on the west coast.*

The dwindling thylacine population was also suffering thanks to its commercial value as a part of the island's wild fauna. Here is just one of very many examples, an 1874 newspaper advertisement: 'Wanted. Tasmanian Native Tigers, Devils, Wombats. Porcupines, and Platypus. All these animals I want dead, and perfect. J. Omant, Furrier, 22, Elizabeth Street.'[17]

Two motions in support of establishing a government bounty were ignored, but its backers persisted, until east coast farmer and House of Assembly member John Lyne succeeded in having a motion tabled. The crucial, final debate on the Bill to eliminate the thylacine from Tasmania—to make it disappear, just as the Aboriginal people of the island had been considered valueless and dangerous and therefore extinguishable—is a peculiar reminder that democracy's strength is also its weakness: it can be so easily abused from within. This was nothing less than an extinction trial, to determine the guilt or otherwise of a species. Following is the ghostly but vivid death sentence, being the Parliamentary proceedings as reported by the Hobart *Mercury*, 5 November 1886:

> TASMANIAN DINGOES. The House then went into committee to consider an address to be presented to His Excellency the Governor praying for the appropriation of £500 for the destruction of tigers, otherwise known as Tasmanian dingoes.
>
> Mr. LYNE in moving the motion, said it might be taken that 100 sheep were destroyed per dingo, and the destruction of 500 dingoes would preserve 50,000 sheep. He quoted some extracts in support of his contention that sheep were decreasing and dingoes increasing. What he would propose was that £1 per head should be paid for every tiger. It was no joke. He reckoned 30,000 or 40,000 sheep were killed annually by dingoes. They ran whole flocks down into gullies, and maimed more than they killed. The scientific name of the animal was *legunta*, and it was one of the greatest pests the colony had.
>
> Mr. GREY supported the resolution, and pointed out that the inhabitants of Buckland and Spring Bay had formed themselves into associations for the destruction of these animals. The animals were allowed to breed on the Crown lands, and for that reason the cost of their destruction should be met by the Crown. In many cases they had become so numerous that Crown lands had been largely given up by the sheepowners, to the loss of the State.
>
> Mr. DAVIES trusted that the hon. member's resolution would be carried, because he had many communications from his constituents about Avoca and St. Paul's [east coast] of the loss they sustained by the ravages of the tigers. He was quite certain that the State would gain much more than this £500 that it was proposed to spend in the destruction of these animals

in the increase of sheep that would take place. He had been informed that sheepowners in the district he represented lost 20 per cent. of their sheep by the tiger, and hoped the resolution would be carried in the interest of the country.

Mr. FENTON opposed the resolution, and ridiculed the idea of £500 doing all the good that was expected of it. He did not believe that the Crown land runs had been given up on account of the ravages of the tiger, and attributed it to the low price of wool inducing sheepowners to give up the poorer runs they rented from the Crown. He contended that the vote could not do any good for the farmers which they could not do for themselves.

Mr. BRADDON said that if the loss was 50,000 sheep a year it meant £50,000 at least, and he could not see why the sheepowners should not spend this £500 themselves to make such a gain. It would pay the sheepowners to spend £15,000 to save themselves from such a loss. He would be very willing to assist the hon. member to quell this *leghunter* [a thinly-veiled reference to Lyne's sexual proclivities], but it was absurd to think it could be done by such a vote. It was wrong of the hon. member to drag in the Crown lands and heap the blame upon them. He should, in the language of the Attorney-General, take a straightforward course, and move a vote of want of confidence in the Government for not having caught the tigers.

Mr. SUTTON supported the motion, not in favour of the sheepowners, but of the small farmers in his district. The Government should assist the efforts of private individuals who were already giving £1 or £1 10s. a head for the tigers in some places, as it would be more than returned in an increased quantity of mutton for the consumer.

Mr. PILLINGER said it could not be expected that private landowners should go on destroying these animals while they were being bred and were multiplying on the Crown lands of the colony. It was not only affecting the lease of Crown lands, but it was preventing the sale of Crown lands to a large extent. Landholders could not afford to pay for the destruction of tigers on Crown land. They were already paying largely for the destruction of the animals on private land. He had been paying £1 a head for dead tigers for years, and if the vote was given at the rate of £1 for £1 spent on the destruction it would greatly assist the landholders and benefit the whole country. He hoped hon. members would realise the fact that the question was one affecting the interest of towns as well as the country, and that the question would not be made a funny one, as the hon. member for West Devon had made it.

Mr. DUMARESQ could not support the vote, as he did not see that the tigers could be any more numerous now than they were formerly. Nor could he understand why they were not destroyed, if they did so much damage, by the sheepowners themselves.

Mr. HAWKES said as the miners of his district had driven these animals out into the hon. member's district by appropriating their natural food, it was only a matter of justice that he should support the hon. member; but he might shortly have to ask for a vote to exterminate another animal his constituents were suffering from—the yellow agony—and he would then ask for the hon. member's support.

Mr. BARRETT: What, the Chinamen?

The MINISTER OF LANDS said that after what had been said by the hon. member for Oatlands there was not very much for him to say. The hon. member for Oatlands said that there were over one million acres rented from the Crown by pastoral tenants, but he forgot to state there were millions of acres unoccupied, which were as much the breeding ground of tigers as the land occupied. He knew that lands were thrown up especially in the Lake districts solely because of the increase of those tigers. He gave particulars of the large amounts paid by owners of lands in getting rid of this pest, and viewing the Government in the light of owners of an extensive and valuable estate he did not think that the vote was out of proportion to what had been paid by others.

Mr. DOBSON hoped that the motion would be amended in order that the money might be spent only for the destruction of tigers upon Crown lands. So many extraordinary statements had been made as to the habits of this animal and its terrible ravages that he thought it would be worth while to refer it to a Select Committee to see what the facts were. Inasmuch, however, as they had granted public money for Californian thistle and rabbits, he did not think that the hon. member was wrong in asking the House to assist in the eradication of this pest.

Mr. W. T. H. BROWN said that the Crown lands were the chief breeding places of the tigers, who seemed to kill the sheep simply for amusement. Very often when a tiger was caught it proved to be a female with cubs, and it seemed as if she had killed the animals to teach her young.

Mr. LYNE, in reply, said he dealt with this matter as one affecting the whole colony, as if the country suffered it would extend to the towns. He asked £500 simply to see how the thing would work. He had a letter from a gentleman who said that he was a poor man, simply because of the destruction caused by those animals bred upon Crown lands.

Question put,—'That the resolution be agreed to,' and the committee divided as follows:

AYES, 12—Messrs. Dodds, Fitzgerald, N. J. Brawn, Lewis, Mugliston, Davies, Sutton, Grey, Pillinger, Lyne, Hawkes, W. T. H. Brown (teller).

NOES, 11—Messrs. Braddon, Conway, Bird, Fenton, Young, Mackenzie, Hart, Hartnoll, Dumaresq, Barrett, and Crisp (teller).

Question resolved in the affirmative.

The resolution was agreed to, and reported, read a first and second time, and agreed to.

The thylacine thus fell, victim of outrageous statistical exaggeration, blatant untruths and the buying of the racist vote. The power and influence of a political group—wealthy landowners—won the day over reasoned consideration. Facts did not matter; they scarcely entered the debate. A strange fledgling democracy was at work, its ineffectiveness highlighted by a rural writer's letter to a newspaper:

> Sir,—I see the honorable member for Glamorgan wants £500 of public money to destroy hyenas (so-called tigers). There are compartively [sic] few sheepowners troubled with those vermin, and it is no affair for Parliament to be troubled with, for it would be a great injustice to the whole country in paying for the benefit of a few wealthy sheepowners. The farmers could more reasonably ask for help to destroy the parrots, which are on the whole much more destructive,—Yours, etc. FARMER.[18]

The Mercury reported the ballot in a manner probably reflecting waning public perception of the animal as a menace:

> The tiger, or dingo, received a very bad character in the assembly yesterday; in fact, there appeared not to be one redeeming point in this animal. It was described as cowardly, as stealing down on the sheep in the night and wantonly killing many more than it could eat, as being worthless even for its skin.[19]

Certainly, being a predator, the thylacine ate sheep. It had been doing so since 1804. Severe encroachment upon its territories may even have forced dislocated populations to target sheep, lacking their natural prey forced away from enclosed lands, and thereby giving an impression of increased thylacine numbers. But that can only ever be speculation. What is beyond doubt is that after 80 years of fitful skirmishing, this symbol of the untamed island had been sentenced to death.

9

CAPTURING TIGERS BY THEIR TALES

'It has the face of a dog, the eyes of an owl, the stripes of a zebra, and a pouch like his cousin, the giddy kangaroo.'

So wrote the Iowa, US, *Davenport Leader* newspaper in 1896, in describing the 'Tasmania Zebra-Wolf, the most destructive and dangerous of the marsupials . . . those animals that carry their babies in their pockets.'[1]

This new chapter records a selection of the many and varied descriptions of the thylacine over time, written in Tasmania, mainland Australia and overseas. Accuracy and exaggeration clash repeatedly, and presumably uncritical acceptance by readers did the animal no favours. Accounts of the thylacine varied so greatly as to reconfirm its unique status as an instantly recognisable yet ultimately unknowable top-order marsupial carnivore—Tasmania's wraith.

Tasmania, 1829:

> In running it bounds in the manner of a kangaroo, though not with such speed . . . The skins are certainly beautiful, being well adapted for saddle cloths, and until the horse guards at White-hall are more decently caparisoned by their help, our

military in Van Diemen's will doubtless adopt them and add to their warlike appearance on field days.[2]

A century later, in preparation for a visit to Cradle Valley's Waldheim Chalet in a new-fangled motor car, this 'caparisoned' observation was made:

> Our driver completed his costume by draping his shoulders with a Tasmanian tiger skin, which usually does duty as a floor mat.[3]

UK, 1845:

This account by eminent Polish scientist Paul Edmund de Strzelecki is a good example of near accuracy based on close personal observation:

> It makes a peculiar noise, sometimes resembling the growl of the common tiger, at others the bleating of a lamb . . . It is said to be stupid and indolent; but this is a mistake . . . When the mother was killed, the young were found to adhere so firmly to the nipple, that it had to be cut, and the mouths of the young were then forced open. They lived about two months in a room, and were then found dead in the fire-place, to which they had, it seems, retired for warmth. A *Thylacinus* that I saw in possession of a gentleman had a puppy [dog] six weeks old thrown to it, which it immediately tore to pieces and devoured.[4]

Mainland Australia, 1866:

> At some remote period, when Tasmania was still part of the Australian mainland, this bloodthirsty animal existed not many hundred miles from Sydney . . .[5]

Mainland Australia, 1867:

'Encounter with a Tiger', Illustrated Adelaide News, *Saturday 23 March 1867, page* 6.

The gentleman whose adventure is pictured in our engraving was looking for horses on the Blackboy Plains, near Fingal, when on passing a flock of sheep at about 300 yards distance he observed an animal which he took to be a dog of the kangaroo breed trotting through the flock. He passed on, the thought striking his mind that the supposed dog had a rather singular appearance, when turning suddenly round the identity of the animal with the native tiger was betrayed to him by the marks on its skin. A hunt of a most exciting kind was immediately improvised, the tiger leading its pursuer through lagoons and timber, and maintaining tremendous speed. After a chase of an hour and a quarter the tiger began to show signs of weariness, and his enemy being enabled to come within flagellating distance, soon brought it to bay by means of a liberal application of the stock-whip. Although by nature a coward, this savage fights with uncommon ferocity when driven to close quarters. In this instance the tiger carried on the combat with amazing fierceness, but human strength and human contrivance were against the brute, which after a quarter of an hour's desperate resistance, fell dead under the blows of its antagonist, administered as depicted in our engraving.[6]

Tasmania, 1870:

The zebra striped tiger-wolf, the king of the [captive] menagerie is in good form and hisses and growls ominously when stirred up, and requested to walk round and show his muscle.[7]

Sheffield, UK, 1875:

The following article, 'At Home with the Zebra Wolf', was written by an 'ardent geologist' and its accurate descriptions of Hobart give way to a remarkable flight of fancy:

What a lovely morning that was upon which I, in company with three other tourists, set out from Hobart Town to ascend the mighty mountain that rises precipitously to a height of nearly five thousand feet, immediately in the rear of the city. Before we had well cleared the suburbs it loomed majestically in front of us above the confusion of lesser hills at its foot, its hoary brow bare of trees, a most unusual thing in Tasmania, and the grandeur of its appearance enhanced by its crown of snow, glittering with rose and opal tints beneath the rays of the hot Southern sun, which illumined on its sombre front a huge colonnade of Titanic greenstone pillars, with a dark fissure here and there . . .

The city (which by the bye is a considerable place) looked very pretty with its handsome town hall, beautiful Catholic cathedral, Gothic convent, numerous church spires, quaint windmills, toy batteries, streets, terraces, and squares. On one side the river stretched away for miles in a gradually lessening and still lessening line of silver, its banks fringed in some places by dark eucalypti forests, in others by park-like pasture lands of emerald brightness . . . Close to the town rose the towers and turrets of Government House amid the green foliage of blue gums, the gay Union Jack floating from its summit, and almost bordering upon its private grounds we could perceive the many tinted foliage of the trees in the Domain, with its trimly-kept parterres and smooth gravel paths . . .

Having enjoyed the prospect on all sides, we turned our attention to the mountain itself, inspected the 'rocking stone', discovered a large black hole which common rumour asserted to be bottomless, and adown which we cast large stones, listening to their hollow reverberations in their rapid descent until they grew too faint to be longer audible. Next we discovered a small deep lake, a strange anomaly on the top of a mountain . . . Being an ardent geologist, I handled my hammer and set off alone in high spirits.

I was at last interrupted in my engrossing occupation by a thunderstorm, but discovering a chasm or deep fissure in a mass of lichen-covered rock close by, I rushed into it for shelter. I had to proceed some distance, however, to find a dry comfortable resting place, and when I had discovered one and was about to sit down, I started back in horror and alarm on noticing four gleaming sparks as of fire glaring at me from out the darkness. My hammer was dropped, and a pistol was in each hand in a minute. They were both small, delicately fashioned, six-chambered revolvers; but I knew that they were desperate fellows at close quarters, and I also knew how to use them. This was all very well had I ordinary foes to encounter; but those terrible eyes—for eyes I felt that they were—gleaming out of the darkness, belonging to I knew not what, almost unmanned me. Did they belong to serpents? Scarcely, for they did not move as though set in the ever-vibrating head of a snake, neither had they the brilliant colouring or the metallic lustre. No; they were only red—red as living coals of fire. 'They must belong to beasts of prey, and one of them must be within a dozen yards of me,' I muttered to myself . . .

Then I heard a savage snarl, like that of an angry dog, but deeper, fiercer, more prolonged and then I guessed what creatures I was opposed to. I concluded that they were Tasmanian tigers; that this cave was their den, their lair; and I knew that as likely as not there were a dozen hidden away in its dark defiles; and if so, the report of my pistol would very probably bring the whole pack down upon me. A pleasant thought this; but as there was no way of getting out of the difficulty except by showing fight, and to do this I must discharge my pistols, come what would of it, I took deliberate aim right between the pair of fiery eyes that were nearest, and was about to pull the trigger when I noticed that the red orbs were slowly approaching me.

'Come on, my dear friend,' I muttered to myself in grim satisfaction; 'the nearer mark, the better shot. I'll pour my little leaden pills down your very throat if you'll only come nigh enough.'

Onwards crept the cowardly slinking brute, closer and closer yet, until at last I could actually see its great head, its expanded jaws, and huge gleaming fangs, and could even feel its hot breath on my face. I saw that it was about to spring, and I fired. The ball must have entered the roof of its mouth and passed through its brain, for it fell back into the darkness without a cry or howl.

Then I saw the other two gleaming points of fire advancing upon me like the red lamps of an express train flashing through a tunnel. I rapidly fired three barrels in succession, but every ball missed, and the next instant I felt myself hurled on to my back, and sharp fangs entering my left shoulder.

Had I lost my nerve at that terrible moment my life would have been terminated then and there; but I knew that my right arm was free, that my right hand still grasped my revolver, which yet held a couple of charges.

I strained my neck to catch a clear view of the animal's head in the gloom. I succeeded, and placing the muzzle of my pistol just inside one of its ears, I fired, and releasing its grip on my shoulder he too fell over on his side, dead.

Springing to my feet, I now glanced round for any more fierce, hungry eyes; but to my great relief, encountered none. I then dragged my two dead antagonists out into the daylight, and discovered them to be zebra wolves, male and female, fully six feet in length to the tip of the tail, and as large as jackals.

You may imagine that I was the hero of the expedition that day, and the victims of my prowess were duly stuffed, and now adorn the Hobart Town Museum.[8]

Mainland Australia, 1879: (Compare and contrast!)

He is possessed of a most formidable set of teeth and remarkably powerful jaws, but is one of the most harmless of carnivorous animals. An old shepherd walking regularly over the run with a couple of sheepdogs is sufficient to keep him away. Even a newly topped-up brush fence has been known to keep him out of a paddock . . .[9]

Tasmania, 1884:

During the day's journey it was my luck to fall in with a man of whom I had heard much as the bearer of H. M. mails between

> the N.W. Coast and Bischoff in the pre-tramway times . . . Clean shaved, long haired, and muscular. Such is William Byrne . . . The night we spent in desultory talk by the fire. My companion lay in his tree munching biscuit crumbs—all he seemed to care for—and told me that during the past 12 years he had travelled roughly 32,000 miles, or something about 100 miles weekly, weighted with mail bags . . . He often flashed his lamp on to half a score tigers and devils feasting on the carcases of horses or cattle bogged in the Nine-mile forest. All nonsense, said he, the tales of tigers one sees in the papers. Even when half-starved, during the winter months, they will leave the half-devoured carrion and sneak into cover at sight of a human being.[10]

UK, 1885:

> There has recently been received at the Zoological Gardens a pair of Tasmanian wolves . . . When embarked, the female had been caught for six months, and had become so tame as to permit herself to be touched through the bars of the cage. The pair were fed almost exclusively on mutton, showing a preference for the parts containing bone; they refused the flesh of a wombat which was offered them when in Tasmania. This partiality for mutton must be an acquired taste—one that could not have existed previous to the advent of the colonists.[11]

Tasmania, 1889:

Demonising the thylacine was common; humanising it all too rare:

> One day recently, as Mr Moone and some others who were assisting him at his cattle station, at the Vale of Belvoir, were wending their way near that place they saw a large tiger

trotting leisurely across the course they were pursuing. Two of the party seemed elated at the prospect of an encounter with the animal. Thinking, no doubt, that the skin would be an admirable trophy, and that it might be fairly claimed by him who should prove himself the bravest of the brave on the occasion, they flew to the front with apparently every demonstration of laudable rivalry. In the heat of the moment, thinking that the assistance of two dogs that were with them would be advantageous, they 'sooled' them on the tiger. This was, however, an outrage which the latter could not brook. Turning with snapping jaws he drove one, a sheepdog, completely away, and finding that the other, a kangaroo dog, would not come to close quarters, he turned his attention to one of his biped aggressors. The latter had, however, on seeing the turn affairs were likely to take, placed himself in close proximity with a convenient tree, which he ascended with marvellous celerity. Tiger next drove his other would-be assailant to a tree, and then catching sight of Mr Moone, who was standing his ground, revolver in hand, advanced to within a few feet of him, and then pausing regarded him for a few seconds with steadfast gaze. Mr Moone did not fire, and tiger, deciding apparently that the gentleman was only acting on the defensive, and with no hostile intention, started at an easy trot in the direction he was travelling when so unceremoniously interrupted.[12]

New York, US, 1894:

The thylacine (sometimes called the Tasmanian wolf) is very fast disappearing, a state of affairs for which the Tasmanians are devoutly thankful . . . This animal is particularly noteworthy for the fact that the male carries its young in a pouch as well as the female.[13]

Illinois, US, 1897:

He is a night prowler. The light of day blinds him as it blinds an owl.

Mainland Australia, 1897:

One of the most noticeable points about it is its enormous eyes, and the fixed and truculent stare . . . Its gait, like that of the dingo, is a sort of loping canter, which it seems to be able to keep up for an indefinite time. It runs on scent like a dog . . .[14]

New York, US, 1903:

Though these animals are said to be very vicious, this particular one is perfectly harmless and wholly indifferent to the presence of man. The keeper enters the cage to treat her wounds when she is feeding, and the only notice she takes of him is, when stung by the application, to walk quietly to the other side of the cage and then return immediately to her food. Her face never assumes the ugly snarl of the wolf . . .[15]

New York, US, 1903:

Australia and Tasmania possess several creatures which have never been tamed . . . The most savage of the lot is the Tasmanian devil . . . Larger and equally savage is the Tasmanian tiger . . . A single thylacine will kill twenty sheep in a night . . . the Tasmanian tiger, if caught young, appears to become quite tame, but as soon as it is full grown its native savagery returns, and if it once tastes raw meat or blood it becomes utterly unmanageable and very dangerous.'[16]

At least one Australian bit back against such nonsense. That was Victor Daley, the journalist and poet who wrote under the pseudonym Creeve Roe. In 1903 Daley responded to a statement made by Carl Hagenbeck, the famed German animal merchant who caught and sold many thousands of animals during his lifetime, and who designed the 'modern' zoo, without the cramped, caged-in enclosures that had characterised earlier menageries. Hagenbeck had also described the thylacine as 'untameable', killing 20 sheep a night, and hardly ever seen. Daley wrote a sarcastic response:

> It is well known to every Australian journalist worth his liquor that the animal referred to is mostly a native of Queensland, but is also found, at times, in New South Wales, and also Balmain and Five Dock. Its stripes are usually plain scarlet, and yellow and blue. Some rare specimens are a little more gaudy, and are fine chromatic studies . . . I am prepared to give anybody who can supply it a guinea for a live Tasmanian tiger, and will then be able to make money out of the deal. But I don't want one that is larger and heavier than a policeman.[17]

Tasmania, 1908:

> The brute has been seen several times in broad daylight . . . The dogs in the neighbourhood are said to be in a state of terrorism, and howl loudly when the 'native' appears in the locality . . . the horse jibbed at sight of the animal and refused to advance towards it . . . its arrival is invariably heralded by the howling of numerous dogs.[18]

Kansas, US, 1909:

> About as untameable an animal as ever came into captivity has been added to the collection in the form of a Tasmanian wolf.

The new arrival, the first of its kind received for a long time, hates mankind with a deathless hatred and makes no pretence of gentler feelings. It lies coiled up in the remotest corner of its den all day, even the offer of food being an excuse for an outburst of boundless fury.[19]

Tasmania, 1915:

The true species is absolutely fearless in man's presence, and if he smells meat will follow the wayfarer on the lonely bush tracks for miles, stopping when he does and going on when he resumes his walk . . . One was shot the other day by Mr Fred Dempster on the southern side of the Arthur River, but it took three bullets to kill the animal. The first two simply glanced off the skull and failed to penetrate the thick covering of muscle that envelopes the head. The brain instead of being in the middle of the forehead, as in other animals, is just over the nose . . .[20]

Pennsylvania, US, 1949:

The Loch Ness Monster—or something—is terrorising the unimaginative folks of Mountain River, on the fringe of Tasmania's 'Unknown Corner', which is the uninhabited extreme of Australia's island State. Except for a posse of 20 men scouring the hills, Mountain River people keep indoors at night because of a 'phantom' animal which makes no tracks yet neighs like a low-throated horse.

Dogs turn and run. Experienced bushman Jack Oates no doubt did likewise when the creature jumped up five feet ahead of him. 'It had a big head and a broad white chest' . . . One theory: the creature is a 'tiger' (*Thylacinus cynocephalus*)—thought by most to be extinct. The tiger was, or is, a carnivorous marsupial like a zebra-striped dog.[21]

Ohio, US, 1960:

The thylacine, a creature thought extinct, has been reported seen in Tasmania. The beast is related to the 'possum, looks like a wolf, jumps like a kangaroo and is partially striped like a zebra. A one-animal zoo?[22]

Pennsylvania, US, 1964:

Q—Which is the largest of the flesh-eating marsupials?
A—The Thylacine, a marsupial wolf. Wiped out in Australia, this savage creature is a making a last-ditch stand in Tasmania.[23]

Arkansas, US, 1980:

Before anyone realised the species was endangered, it had all but disappeared.[24]

Tasmania, 1982:

The observer was asleep in the back of his vehicle. Light rain was falling at 2.00 am when he awoke and, out of habit, scanned the surrounds with a spotlight. At the extreme end of his arc he saw a large thylacine standing side on 6–7 m distant. His camera bag was out of immediate reach so he decided to examine the animal carefully before risking movement. It was an adult male in excellent physical condition and had 12 black stripes on a sandy coat. Eyeshine was pale yellow in the beam of the spotlight. It moved only once, an enormous yawn in the best thylacine tradition, probably a displacement activity. After several minutes the observer attempted to reach his camera bag but in doing so lost the animal from the beam. He recovered it as it moved away into the undergrowth. Leaving the vehicle and moving to where the animal disappeared, he noted a strong, musky scent. Despite searching stark naked, camera in hand, no further trace could be found. It is probable that the animal was travelling along the road, came across the car and was investigating when the observer awoke. Like most wildlife disturbed by a spotlight, it eventually moved off into the bush. An all-night search failed to relocate the animal or find traces of it.[25]

10

'THEM BLOODY USELESS THINGS' 1888–1930

> We own a property called Tiger Hill. The original owners were the French family, and it was my privilege to have met & had long conversations with Mr Ab. French. His brother had hold of a Tiger by the tail, & Mr Ab. French clearly remembers taking a possum snare from his pocket, slipping the noose over the animal's jaws, then both dragging it to an enclosure near the house. The Tiger was kept for five months, till a traveller called and offered the boys five pounds for it. In the year 1914 five pounds was a lot of money, so the Tasmanian Tiger was sold.
>
> William L. Ferrar

> They were captured a short distance from Bridport in a snare on the top of a fence, up to which a drag was laid. Messrs. Brewer, who captured them, state that out of 60 caught these were the only pair taken alive.
>
> *Launceston Examiner,* Friday 7 September 1883, p. 2

Uncertainty, confusion and misinformation: to this list of strikes against the thylacine may now be added irony, for even as the government bounty took effect and profitable slaughtering started in earnest, live specimens were also acquiring value. Thylacines had been sporadically displayed in public

A thylacine pair in the National Zoological Park, Washington, DC. They appear to be in remarkably good condition, given their long sea voyage from Australia. Thylacines were also bought by zoos in New York, Antwerp, Berlin, Cologne, Paris, Madras and London. (Smithsonian Institution Archives)

There is a marked size difference between this male and female at Hobart Zoo. A total of 45 thylacines were displayed in the zoos in Hobart and 76 in Launceston. (Collection Tasmanian Museum and Art Gallery)

and private collections in Tasmania and elsewhere since 1817, but demand increased as zoos flourished. (The five pounds mentioned above is equivalent to over $600 in 2023.)

It did not take long for intrigue around the thylacine to grow. As early as 1851, at a monthly general meeting of the Zoological Society of London:

> Upwards of fifty species of animals which had not been previously exhibited, have been added to the collection; and among them [are] three of the most singular and interesting in existence, namely, the hippopotamus, the wart hog, and the thylacine.[1]

In adding thylacines to their collections, zoos helped to slowly shift public perception of the animal. And its confirmed rarity meant that some urban Tasmanians, at least, began to cautiously reappraise their strange carnivore's terrible image: 'On Saturday last two fine specimens of the Tasmanian tiger were brought to town and lodged in the garden attached to the caretaker's house, Town Park. They were captured by a young man in the vicinity of Bridport. Many visitors have been to see them, and admired them greatly'.[2]

By the late nineteenth century, European zoos had become important centres of scientific research and, to an extent, public education. They were also potent flagwavers for the glories of imperialism, holding in their cages exhibits from the far-flung ends of the conquered world. Distant Tasmania's marsupial wolves were as exotic as any other creature, even though they apparently lacked charisma behind wire.

As noted, thylacines were shipped to numerous zoos overseas, a known total of 40 being displayed. In Australia an estimated 184 were displayed in Adelaide, Melbourne, Launceston, Hobart and Sydney's Taronga Park and Moore Park. They were keenly sought by institutions, particularly when extinction warnings began to be made and it was also realised that they did not breed in captivity.

Some zoos may have had a sound motive in wanting to understand and even somehow preserve exotic animals, but cruelty all too often prevailed. Annis Hardcastle Knight (mentioned in Chapter 4), made this observation in 1902:

> The Washington Zoo possesses three [thylacines], a mother and two babies, which arrived some weeks ago in very poor condition as the result of a month's voyage from Australia in the confined limits of a box too small for the adult even to turn around in. Consequently, her sides were badly rubbed and her legs were cramped. She lay listless the greater part of the time, and, when she did move, appeared to do so with difficulty. The little ones recuperated more quickly than their mother, and tried to engage her in their sport by jumping upon her back and rolling down her sides. She is now beginning to take a livelier interest in things, and will occasionally run with the cubs, and leap in the air much as a kangaroo would do.[3]

As the century drew to a close, there arose evidence of a new Tasmanian pride in being able to offer up rather than only put down the thylacine: 'A fine specimen of the Tasmanian tiger is to be seen on board the Barque *Ethel* . . . The *Ethel* will maintain the reputation of her predecessor in taking home Tasmanian animals to the Old Country'.[4] This 1891 report reflects a noticeably different attitude to that insisting on their obliteration.

While Tasmanian loyalty to and love for the Crown remained very strong, as action in the 1899–1902 Boer War was soon to show, the desire was no longer to replicate Merrie England in the Southern Hemisphere. Instead, exports to the mother country showed that love: fruit, fine jams, soldiers and, amazingly, thylacines. Prices offered for them rose steadily. The absurd situation came to bear in which the Tasmanian government bounty scheme, over a twenty-one year period, steadily paid £1 per dead thylacine while zoos offered greater and greater sums for increasingly rare live specimens. The last £1 bounty was paid in 1909. The export price at that time was about £7 or £8. In 1926, London Zoo purchased its last thylacine for no less than £150.[5]

> AN UNWELCOME VISITOR. A short while ago several sheep were found dead in this district [Tyenna], and the mischief was put down to dogs, although none could be caught. One morning Mr. Quarrell, being on the road at daybreak, was surprised to see a tiger a short distance ahead. He immediately went to the attack, armed only with a stick, and aided by his dog. Mr. Quarrell's object was to get the tiger alive, as the price obtainable for a specimen was a consideration. He managed to get a hold, and to retain it for some considerable time, but unfortunately his cries for assistance were unheard, and he was obliged to let go. The tiger made off into the scrub, followed by the dog, but he was not to get away, as Mr. Quarrell obtained assistance, and followed. After a rough scramble the fight was renewed. Unfortunately, it was found necessary to kill the tiger, which was a very large one. The sheep mystery was thus cleared, and a subscription started among the sheepowners was well contributed to. Fortunately, Mr. Quarrell came out of the contest with only a bitten foot, but with a good deal of the tiger's blood on him to bear witness that the struggle had been a hard one'.[6]

Mr Quarrell's object—to get the tiger alive—is a sure indication that this new post-bounty financial incentive had become common knowledge, and this would remain the case well into the future. One wonders how many thylacines suffered the same fate, killed trying to avoid capture. Lacking tranquillisers and other humane and safe ways of securing wild animals, the trappers of those times can't be blamed for their actions. First their government, and then international zoological institutions, called upon their skills to deliver thylacines—dead, but then alive. It was a strange and sadly late turning point in the animal's history.

The dreaded Tasmanian wolf had therefore come to profit numerous groups: snarers and hunters through the bounty and the sale of pelts, exporters, zoos through their admission fees, and the sheep farmers twice over, for the thylacine had become a convenient cover for stock losses through mismanagement, and the government, not the farmers, was stumping up the bounty fees. The irony of all of this has yet to peak, with the costs associated with cloning a single thylacine sure to reach astronomical levels.

That pleasure could be taken in the concept of saving a thylacine, as suggested in the report of the *Ethel* taking one 'home', is an indicator of a change taking place in the collective psyche of the population. At the end of the nineteenth century, there were many new preoccupations to drive that change. The frontier mentality associated with the taming of old Van Diemen's Land had become history as industries such as mining, dairying and timber began to power the economy and create a modern society. Tasmania's population was always small but in the decade of the 1890s it rose from just under 115 000 to just under 186 000: a manifold leap. Yet it was claimed that even greater progress and prosperity were being held back because 'the best agricultural land remained locked up in sheep and cattle runs . . . these owners were driving the industrious settler away'.[7]

This put new pressure on the heretofore all-powerful rural lobby. The island no longer needed to ride on the sheep's back.

Furthermore, could the unthinkable now be thought? Might that curse of the wealthy sheep farmer, the Tasmanian tiger, actually have sympathy extended towards it and be seen as worth saving? Apparently:

> A SICK TIGER.—The Adelaide *Register* says:—Latest advices from the sick chamber of the ailing tiger at the Zoo state that the patient is as well as can be expected. Veterinary-Surgeon Bickford lanced the bad foot, and the poor beast was much relieved. It had suffered so much pain that its ferocity was quite gone, and it submitted quietly to the relieving operation. The head keeper of the Zoo has been a kind nurse to the sick tiger, and the animal has recognised a friend in him, allowing him to constantly bathe the swollen foot with hot water and rub it with ointment prescribed by Mr. Bickford. This tiger is a valuable animal, and worth all the attention bestowed upon it.[8]

It is to be assumed that the word 'valuable' was applied by the journalist in a monetary context, referring to moneys originally paid by and then received by the Adelaide Zoo, but there is in that final sentence something approaching genuine acknowledgement of its true worth.

Naturalists were also out and about, and their published thoughts and beliefs went against the grain of common thought, insofar as they called for a respect for a Nature which ought not to be dominated—implying, partly, a continuing fear suppressed through force—but admired. This was radical, harnessing the Romanticism of a century ago to the burgeoning scientific power and certainty of the fossil-fuelled, electric new world (Launceston was the first city in the Southern Hemisphere to switch on, in 1895, its hydro-electricity generated by the Duck Reach Power Station). The naturalist Reverend Henry Dresser Atkinson, an Englishman who fell in love with Tasmania, who was Truganini's friend and has real claim as an early Tasmanian environmentalist, included this turn-of-the-century 'conversation' in his diary:

'. . . Really I have little sympathy with some of the modern principles of utility, through which so many "things of beauty" in the animal and vegetable world are destroyed to make room for man and his works, or to minister to his appetites and luxuries. Think of the extermination of rare and beautiful hummingbirds to provide ladies' hat ornaments. Ah, my friend, it is a sad business!'

'Then I suppose you would discontinue the bounty on the destruction of Tasmanian tigers?'

'Certainly I would: the *Thylacinus cynocephalus* is a most interesting creature; it is found on no other spot of the earth's surface except here in Tasmania. It used to occur in New South Wales, but is now extinct there. You don't know what an instructive animal it is to the comparative anatomist and zoologist, no, I would let *Thylacinus* have a sheep now and then if he wanted it.'

'What, if you were a sheep farmer yourself?'

'Yes, even then.'[9]

The naturalist Geoffrey Smith took a hardier view, appreciative of the 'grandeur of untouched nature' that was the west coast and cognisant of the meaning of extinction, but seemingly not too fussed about either. He wrote pragmatically:

We could not reach Lake St Clair that evening, so we stopped the night with a shepherd, named David Temple, who made us very comfortable in his wooden cottage. This little cottage, with currant bushes and a few English flowers in the garden, nestling in an Alpine valley and surrounded by the silent gum forests,

This three-dimensional topographic map of Australia's south-east ocean floor shows clearly the huge Bassian Canyon cutting across the continental shelf. Its flooding, 12 000 years ago, isolated the Tasmanian thylacine population and saved it from the circumstances that led to the mainland extinction.
(Geoscience Australia, CSIRO and the National Oceans Office)

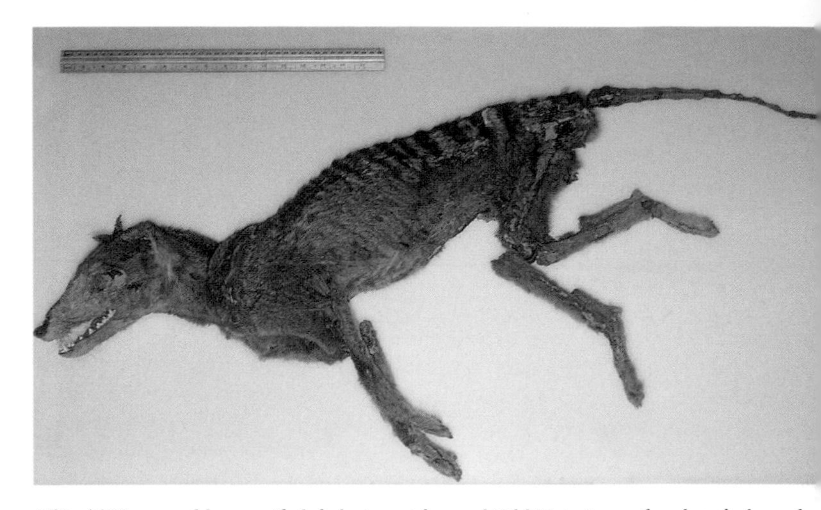

This 4600-year-old mummified thylacine, nicknamed 'Old Hairy', was found at the base of a shaft on the arid Nullarbor Plain in 1966. Mainland thylacines were well on the way to extinction. (Western Australian Museum)

The first known European depiction of a thylacine was by Van Diemen's Land Surveyor-General George Harris. He sent it, together with a written description of the animal, to the Linnaean Soci[ety] in London in 1806. (Gerard Willems, Hobart)

According to Eric Guiler, this is the 'first representation of the Tasmanian tiger in a German treatise of natural history ... It is obvious that the artist had no model at his disposal' (Tasmanian Tiger: A Lesson to be Learnt). (Gerard Willems, Hobart)

his c. 1840 depiction of the thylacine by an unknown European artist is anatomically accurate but a good early example of popular interest in the animal. Gerard Willems, Hobart)

'Thylacinus cynocephalus' *by Henry Richter. Perhaps the most well-known depiction of the animal, it was painted for John Gould and appeared in his* Mammals of Australia *(1863).* (Collection Tasmanian Museum and Art Gallery)

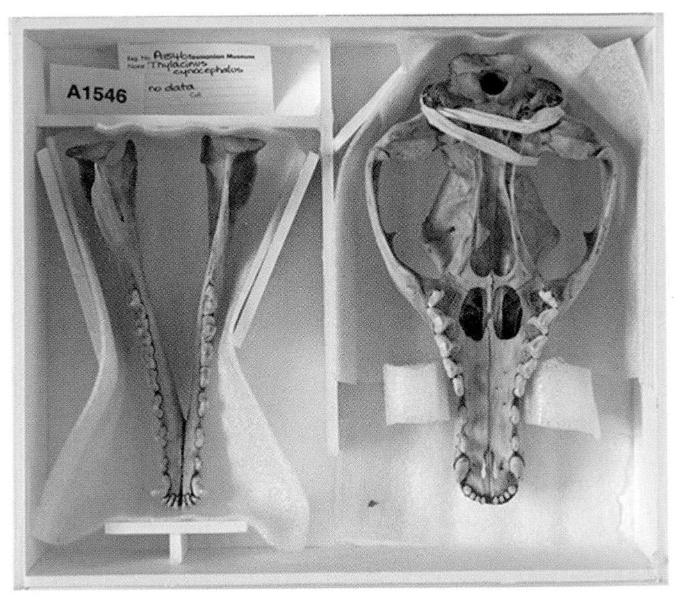

Part of the disarticulated skeleton of the last-known thylacine, rediscovered in the Tasmanian Museum and Art Gallery's collection in 2022.

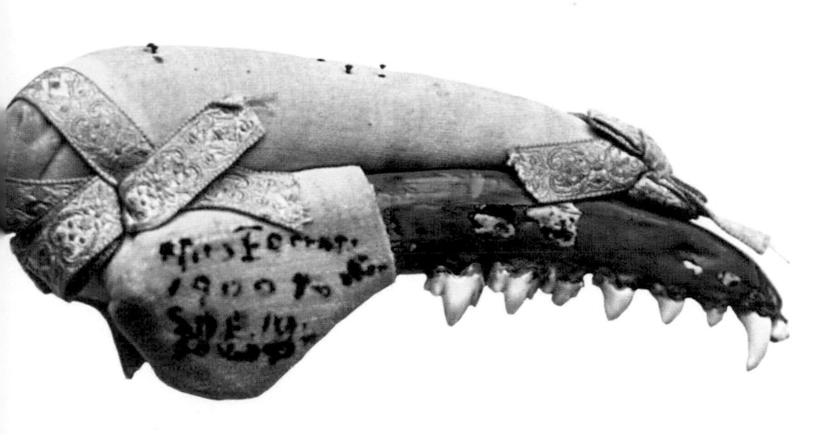

e are very few thylacine artefacts in existence. This pin cushion made from the jawbone of lacine (top), which won second prize in the handicraft section of the Glamorgan Show in), is one. Another is this unique thylacine skin carriage rug (bottom), circa 1903, which was ructed from eight pelts stitched together. (Collection Tasmanian Museum and Art Gallery)

Among the most famous of all thylacine images, this one, taken in 1869 and titled Native Tige of Tasmania Shot by Weaver, *is one of very few 19th-century photographs known to exist. An unusually large albumin print made from a wet-plate negative, it was created through a complicated and time-sensitive developing process. It was possibly taken by Victor Albert Prout, the inventor of the panoramic camera, who visited Australia 1866–1874.* (Tasmanian Museu and Art Gallery)

aren Casey's Got the Bastard *(1991) drew inspiration from the photo* Native Tiger of asmania Shot by Weaver, *1869 (left). It illustrates a hogtied thylacine and its hunter posing ith his bounty. The image epitomises the steady eradication of a species from its natural habitat European settlers. The parallels between the thylacine and Palawa peoples are highlighted rough Casey's imagery and the title of the work, with both persecuted, and the thylacine ndered extinct.* (Oil and mixed media on canvas, National Gallery of Australia)

One of four preserved pouch young owned by the Tasmanian Museum and Art Gallery. This specimen is about three months old, with hair just beginning to grow. In 2002 it was estimated to be worth at least $2 million.

> appeared to me an idyllic place, but the shepherd, who had lived there with his wife for about thirty years, complained of its loneliness . . . He told me many stories of the Thylacine or Native Tiger, which is more abundant here than in any other part of the island, and takes a considerable yearly toll from the flocks of sheep. Since this carnivorous Marsupial is regularly hunted and trapped by the shepherds, and since it occurs only in the little island of Tasmania, it will not be very long before it becomes extinct, so that I was careful to gain any information I could with regard to its habits . . . The shepherds wage incessant war on the creature, in the summer laying traps and hunting it with dogs, in the winter following up its tracks through the snow. A reward of a pound is given for the head by the Government, but the shepherd generally rides round with the head to several sheep-owners in the district, and takes toll from them all before depositing it at the police station. In consequence a large reward must be offered for the carcass of a Tiger, and an offer of £10 during a year for a live Tiger to be delivered in Launceston was unsuccessful. It pays the shepherd very much better just to hack off its head and take it round on his rides'.[10]

The new century ushered in a brave new country, the Commonwealth of Australia, in which the colonies, after ten years of discussion, joined together in 1901. A Tasmanian, Andrew Inglis Clarke, was the chief architect of the constitution. Tasmanians voted overwhelmingly to federate. The minority against were the rural electorates of the east coast and midlands, the electorates which had spawned the lobby groups that pushed the thylacine bounty scheme through parliament. Perhaps this was another indication of how out of step they were.

Alas there was to be no magical cessation of the bounty scheme through a collectively enlightened attitude. That would only happen when it was far too late; a historical embarrassment which would shape Tasmanian politics, and influence global environmental politics through the twentieth century and to the present. Back then, the bounty was good money for snarers, who presented the dead animal to a police station for their £1, or ten shillings if it was a pup. The toes or ears were removed to prevent the snarer taking the body elsewhere for another payment. During the bounty period, from 1888 to 1909, payments were made for a total of 2206[11] carcasses. That, however, may only have been the tip of an iceberg. Echoing naturalist Geoffrey Smith's remarks, Guiler noted: 'Many trappers told me that up to half the thylacines killed were not submitted for bounty but were carted around the local property owners who paid a reward (usually £1) and when the carcase became too smelly it was dumped in the bush'.[12] Furthermore, 'From 1878 to 1896, 3482 Tiger skins were despatched from [a tannery] to London where they were made into waistcoats'.[13] This bold claim, printed in the *Mercury* newspaper in 1968, has never been proven. The poor condition of any number of carcasses, either as the result of the activities of scavengers, or the thylacine's efforts to break out of the snare, would have rendered them valueless. Thylacines may also have eaten the poison baits set by snarers for the scavenging Tasmanian devils. It all suggests a wantonly high kill rate.

The Van Diemen's Land Company continued with its own bounty scheme, payments being recorded right up to the beginning of World War I. Before then, however, the already weakened thylacine population may have fallen victim to disease, as already mentioned. In 1902, a total of 130[14] thylacines were presented for bounty, the annual kill total having been relatively constant for fifteen years. This dropped dramatically to 46 in 1906, to just two in 1909 and none thereafter. The sudden huge decrease across the island may signify a reduction in numbers other than by human agency, although David Pemberton disagrees with the disease theory.

The H. D. Baker story is an intriguing addition to thylacine folklore, previously untold. He was the first United States Consul to Tasmania. In 1909, eminent Hobart residents hosted a farewell dinner on the eve of his departure for a visit to the US. At that dinner, Baker gave a speech:

> I am going home to America, on leave of absence, and I certainly would feel, very much depressed, had I any occasion to think that I would be unable to return here again. Barring any unforeseen accidents, I hope to be back in Tasmania within a few months. I had a love for Tasmania at first sight . . . I feel great regret that the people of Tasmania have not, until recently, seen the desirability of protecting the interesting animal and bird life of the State from extinction. I think Tasmania owes it to the whole world to conserve the rare species of birds and animals which are only found in this part of the world from ruthless destruction. I believe there is a very large number of American and European tourists who would like to visit Tasmania, if only for the purpose of seeing some of the curious animals which they have read about, and which appeal so strongly to their imagination; but it is unfortunate, to say the least, if they come out here, and find that all this interesting natural history—is only ancient history, and will have to content themselves with seeing the bones and skeletons of such animals, instead of the animals themselves, and be obliged to go to the museums to see such relics, instead of being able to see the living creatures themselves in the bush . . .[15]

Some time after Baker's departure, as reported by Hobart's *Mercury* newspaper:

On board the *Marama*. The American Consul's Zoo

A letter received from Honolulu by a friend of Mr. H. D. Baker, American Consul, states that all the Tasmanian animals which he is taking to the United States have crossed the tropics in good health, and there seems every prospect of their reaching their destination in safety. An Australian dingo, which was added to Mr. Baker's collection at Sydney, was quite tame, and would be taken for a walk on deck twice a day. It made many friends among the passengers of the *Marama*. The four Tasmanian devils showed a great thirst while in the tropics, and it was difficult to keep them supplied with drinking water, as in their rush for the water can, whenever the supply was renewed, they would fight with each other, and usually spill the contents. It was amusing to throw two chicken bones to the four devils, as two tugs of war would then ensue, and the devils would pull desperately against each other for possession of the bones. These tugs of war would usually end in draws, the bones becoming broken, and each devil being obliged thereby to exchange his community interest in the bones for actual possession of a particular part.

The devils seemed to lose much of their savage demeanour towards outsiders as the voyage progressed. At first, if a stick was poked into their cage, they would open their mouths and snarl and then snap savagely, but after a while it was impossible to get a growl out of them, and they would playfully gnaw at anything stuck into their cage. It seemed as if the general opinion that the Tasmanian devil is impossible of taming might be incorrect.

The opossums showed a remarkable taste for bananas, and an entire banana would be devoured by one of these animals with the same glutinous avidity with which a cat will eat a fish.

The five wallabies which Mr. Baker took with him appeared to thrive on a diet of lucerne and bran, and they also seemed to

> enjoy a little apple or banana. The kangaroo rat would eat almost anything, but apples and pears appeared to appeal most to his palate. The butcher birds and plovers would eat only chopped meat. The black swans, which were presented to Mr. Baker by [Tasmanian Governor] Sir Gerald Strickland, were much admired by the passengers of the *Marama*. A large filter of water was placed just outside their cage, and their great joy in life was to stretch their necks out of their cage and shampoo their heads in this water. They drank a great deal of it, but their chief use for it was to keep their heads cool by constant bathing.
>
> Owing to the large family with him, Mr. Baker decided not to stop at Honolulu, as at first intended, but to go direct to Vancouver. The animals are to be placed in the Vancouver Zoo, while Mr. Baker pays Seattle a visit, and then will travel on to their destination, some being given to Lincoln Park, Chicago, and some to the National Zoo at New York.[16]

And yet this appeared, in a lone US newspaper:

> H. D. Baker, American Consul at Hobart, Tasmania, who was a passenger on the *Marama,* brought with him a small sized menagerie including two Tasmanian tigers, a black swan, a dingo and other animals and birds for the Yosemite National Park.[17]

Why would one newspaper make such a report? Did H. D. Baker's passion for Tasmanian wildlife lead him to smuggle two thylacines into Yosemite? However unlikely, the mere fact of the reportage is as tantalising as the ageless Tasmanian tiger question: *are they still out there?*

In 1914 the professor of biology at the University of Tasmania, Professor Thomas T. Flynn, called for a sanctuary to be established somewhere in Tasmania to save the thylacine.

His was not the first contemporary voice to raise the alarm, but his academic standing, the fact that he was a Tasmanian rather than an eminent outsider, and that he was calling for a reserve, places him in a notable position in the history of the Tasmanian conservation movement. (That he was the father of the actor Errol Flynn must remain a mere footnote in the thylacine story but not the fact that he later embarked on some questionable deals selling thylacines and other wildlife.)

Then as now, however, an urban academic calling for rural change could not readily expect that change.

The naturalist's view (let the thylacine take the odd sheep) and that of the academic (create reserves to save them) were well ahead of public opinion in the critical early years of the twentieth century, when the bounty and disease together were hastening the destruction of the species. There was then no understanding of, nor a desire to understand, the value of wild animals as parts of complex ecosystems. Why should there be? To use a contemporary analogy, snakes and spiders are routinely killed for being 'dangerous'. That they inhabit urban areas and eat rats, mosquitoes and flies, all of which cause human illness, is apparently beside the point. Spiders and snakes induce a primal fear—and from that day in 1804 when Knopwood's escaped convicts saw a 'large tyger' in the dense bush of their alien environment, the island's people had much incentive to fear and loathe a dangerous animal.

Effectively breaking down that attitude would not happen while a government bounty remained in place. But more and more voices in favour of protection began to be heard and the public became at least prepared to think about it, rather than be horrified or offended. After all, war had been waged against the creature for a hundred years; in defeat it could be safely viewed in zoos—even safely admired.

An editorial in a January 1909 edition of Hobart's *Daily Post* newspaper made a powerful early call for protection:

> The announcement that the Ulimaroa will next week take from here two marsupial wolves—Tasmanian tigers—for the London Zoological Gardens and a brood of Tasmanian devils for the gardens at Melbourne again directs attention to the need of a sanctuary where the native fauna may shelter and be saved from that extinction which must otherwise be its fate. So rare are marsupial wolves that few people have seen them, and there are probably not half-a-dozen in captivity the world over. Yet little is done to protect [thylacines] here and before many years have passed they may be as extinct as the dodo . . . Tasmania's rare animals . . . are being exterminated by the pot hunter and the new enemies and conditions which come in with settlement . . . it is matter of common knowledge that protected birds are shot all the year round by so-called sportsmen who are happy so long as they can kill something . . . though perhaps the average man will fail to see why, for instance, marsupial wolves should be saved from extinction the reasons are none the less potent, and the country which has lost its fauna has gaps in its history that can never be filled . . . The marsupial wolves which are in captivity at Beaumaris, we are told, were only obtained at considerable trouble and expense, and this shows how rare the animal has become. There is, therefore, much need for early action, and in scientific interests it is to be hoped that it will be at once taken . . . and that end can only be attained by the establishment of sanctuaries . . . Once in the reserve an animal is sacred.[18]

The Tasmanian Field Naturalist's Club, formed in 1904, became the first official group to actively promote conservation. At a later meeting in Launceston, 'There was a consensus of opinion that national parks should be reserved in different parts of the State for the preservation of native fauna. A general committee was appointed to take such steps as may be necessary to cause a bill to be passed through Parliament . . .'[19]

The theory of protecting thylacines in this way was a sound one. Within a year, however, the government bounty had come to a de facto end because there appeared to be none left to kill. Were they already extinct? Not quite: three were presented for Van Dieman's Land Company bounties at Woolnorth in 1914. The following year saw the creation of the Tasmanian Scenery Preservation Board, which set up the first of the island's reserves and the 11 000 acre (4455 hectare) Russell Falls–Mount Field National Park, an hour north-west of Hobart. This park contained some of the world's tallest flowering trees, the mountain ash (*Eucalyptus regnans*).

At this point, therefore, a fundamental shift had occurred. There were no longer any bounty schemes, safe havens were becoming available, and the public had begun to accept that the thylacine might be an asset, not a menace. Now, perhaps, science rather than politics might begin to chart a future for the seriously endangered animal. But in the bush, snaring and hunting went on, for skins and zoos. Thylacines were not to receive full protection for many years.

Snaring had long been an important part of the rural fabric:

> Not only was it a cheap, simple and reliable method of providing meat and skins, it also controlled animal populations eager to devour emerging crops. In the financially precarious years of clearing the forest to form a farm, wallaby meat often fed the family and skins constituted the only cash flow. Snaring was a practice with which most farmers and rural workers were familiar . . . With the European markets putting a premium on the pelts of wallaby and possum, men whose homes fringed the

> high country—farmers, timber workers, shearers, prospectors—would drop whatever they were doing during the slack winter months, and head to the mountains where the best skins were.[20]

That meant a lot of snaring—up to a million possums, wallabies and pademelons annually, sometimes many more—and thylacines were, of course, valuable financial catches. It will never be known how many were killed this way, even after their rarity had become a matter of genuine public concern. Snaring was a legally controlled primary industry with open and closed hunting seasons, but:

> This control of the seasons did nothing to encourage the snarers to operate within the confines of the law . . . With a skin being valuable whether the season was open or not, the vast number of snarers kept the skin. To evade the attention of the police who often raided even the most isolated runs, most snarers had their special hiding places for their out-of-season skins. Hollow trees, clefts in rocks, false panels in huts—all were used.[21]

> How snaring worked:

> The snares require to be strong, about 12 strands of good twine, and the simplest and best to set for them is the neck one, which requires to be wired with snaring wire sufficiently stiff enough to keep the snare suspended by a peg about 18 or 20 in., driven in the ground, and a slit in top to receive the snare, which should be set about 9 in. from the ground, and the snare set 40 in. in circumference. [A man] once took charge of a run many years ago on the south of Mount Cameron, and the tigers were making great havoc amongst the sheep, but by the above means the run was free from the pests, having killed 15 in about three months.[22]

Such a culture of ignoring and evading early faunal protection measures suggests a lack of care for the thylacine's particular future in those who were the last to interact with it in the wild. 'Them bloody useless things' was a comment offered to Eric Guiler by H. Pearce of Derwent Bridge, a member of a family which collectively took many thylacines.[23]

Furthermore, misidentification of the thylacine as a sheep predator continued. In 1906 a journalist investigating thylacine and Tasmanian devil captures for zoos observed that: 'Native tigers have a great partiality for mutton, apparently. It is a common custom to attribute losses of sheep that cannot be otherwise accounted for to deaths by tigers.'[24] This echoes an earlier myth-enhancing observation that: 'Its presence is only indicated by the disappearance or killing of sheep.'[25]

And the following was written in 1922 as part of a lengthy *Mercury* newspaper article about Tasmanian shepherds, the brief extract notable for its internal contradictions as to the identity of the real villain:

> [The shepherd] must also be a skilled zoologist or else he would never be able to locate the lair of a Tasmanian tiger—which animal, fortunately, is becoming scarcer every year owing to progress of civilisation—that might be preying on the flock without his actually seeing the animal at work. Moreover the depredations of dogs given to sheep killing are often a supreme test of his powers in this direction.[26]

Hunting as a sporting activity would not have done the thylacine's cause any good either. The desire to kill animals other than for food, clothing or profit has always been labelled a sport by its practitioners. Thylacines must have rated at the very top of the hunter's list, along with wedge-tailed eagles. There can be no telling how many became the victim of sportsmen, although

'One record of 1910 indicates that there was still much "fun" to be had knocking off Tigers'.[27]

Along with Thomas Flynn, Clive Lord, the director of Hobart's Tasmaniam Museum, played an important role in attempting to save the thylacine. In 1917 he published a paper describing the island's mammals. He noted that it was only the fourth such compilation and that the previous three were not only incomplete but out of print. What was worse, he had had to rely extensively on British Museum catalogues since no local reference collection existed. This indicates all too clearly the almost total lack of official interest in the state's natural heritage or, put another way, demonstrates the difficulties that Lord and like-minded colleagues faced in their attempts to highlight the plight of the thylacine. He wrote:

> It is remarkable that more attention has not been paid to the Mammals of Tasmania. Many interesting species occur here, and as they are decreasing in number, the following notes have been compiled in order to attempt to arouse interest ere it is too late. Unless unforeseen circumstances occur, many of our Mammals will share the fate of the Tasmanian Emu . . . owing to their partiality for killing sheep, a war of extermination has been waged against the 'Tigers'. They are now rare, and are only found in the most rugged parts of the island . . .[28]

Ten years later, and to his credit, Lord was able to claim that that massive indifference had been overcome. But his prognosis for the thylacine was gloomier still. Most of the entries in his updated paper, 'Existing Tasmanian Mammals', are short paragraphs; that for the thylacine runs to nearly two pages. It reads in part:

> Within the last few years considerable attention has been directed to the present state of the Australasian marsupial fauna. The

> meetings held during the recent scientific congresses in Australia have served to stress the importance of our native animals . . . Our fauna consists, to a very large extent, of archaic types, which, when brought into sudden contact with more advanced forms, rapidly decline . . . Australian Zoologists might well consider plans whereby at least an outline biological survey of our indigenous fauna could be attempted . . . [the Thylacine] is now also being killed out even in the rugged and more inaccessible parts of the country, which tends to reduce still further the remnants of this species. The explanation of this is that the Thylacine interferes with the trappers' snares. As a result, a powerful 'springer' snare is set often in the vicinity of their 'skinning yards', which are situated every quarter of a mile or so along the lines of the snares. Thylacines or other animals caught in these powerful snares are, as a rule, too severely injured to be kept alive as specimens for zoological gardens, even if the trappers would take the trouble to bring them in . . . Its eventual doom seems apparent unless such attempts as are being made . . . to breed these animals in captivity are successful . . . It is doubtful if the shy animal will breed within the confines of a Zoo . . . During the breeding season a male Thylacine has even been known to follow the same route across many miles of country, and one particular animal used to regularly leave a trail of slaughtered sheep along the same line of march each year, but he was trapped eventually. If a Thylacine kills a sheep it will usually only suck the blood, and may also take a little of the kidney fat.[29]

Another influential scientific publication at this time, by A. S. Le Souef and Harry Burrell, drew directly upon first-person layman accounts in building up its descriptions of the thylacine. It was less inclined towards the possibility of extinction:

> This animal is now getting rare in Tasmania. The inhabitants seem to have a superstitious dread of the 'hyaena', as they

> sometimes call it, and will kill the wolf whenever opportunity offers. Indeed, some will even smash the wolf to pulp afterward, thus depriving science of the skeleton and skin.[30]

These two descriptions presented a gloomy picture of a wildly mischaracterised animal heading for extinction. There were of course admirers of the thylacine: 'About the size of a wolf, [it] was lying down in its cage, and looked a very graceful object, with its pretty black stripes on its brown woolly fur, its head like that of an intelligent dog, short pointed ears, slender tail, and gazelle-like eyes . . .'[31] But alas such sentiments were well in the minority. Was this the same animal?

> The Tasmanian wolf is one of the most stupid animals; its lack of intelligence is the cause of its untameableness. It never loses its ferocity. Animals are 'wild' for two reasons: (1) Because they are intelligent and know what they dread; and (2) because they are stupid and do not know what they fear: the thylacine ranks very low down in the latter category.[32]

Common decency and rationality—and anger—would eventually outrun such wilful idiocy, albeit far too late. At the 1922 Congress of the Royal Australian Ornithological Union held in Adelaide, a South Australian delegate lamented:

> The frightful mortality and brutality to Australian birds when sending them overseas should be stopped. Zoological societies should not be given control, because they had entirely put themselves out of court . . . [The delegate] had seen things in Sydney which had absolutely appalled him. Zoological societies had formed a combine to exploit Australian fauna. That was most disgraceful . . . Other countries were ready to offer high prices for the Tasmanian devil and Tasmanian tiger. Such offers were an invitation to murder.[33]

Ultimately, it was through Lord's efforts that, at last, there came a form of protection, in 1930. But how protective was it? Now, thylacines could not be hunted in December, for it was believed that they bred during that month. From today's perspective it seems a pathetic gesture but at the time it was a hard-won political victory.

Even so, public sentiment held firm. In the debate about the changes to Tasmania's game laws, a 1928 letter writer to the *Mercury*, signing himself 'On The Land', declared that: 'The suggestion to protect the Tasmanian tiger in its natural state has nothing to commend it. This indigenous animal beyond being curious, has nothing to recommend its perpetuation.'[34] And a letter writer some years later, a 'Robt. Stevenson' in 1937, went further still, proudly:

> I think I can claim to have had as much experience of the 'animal pest' as any man in the state today . . . Many years ago after doing a hard day's work I would take my gun, a sheep dog, and an empty wool pack, for sleeping equipment, and camp below the sheep's bedding hill so that when the tiger chased the sheep I could run in front and get a shot at it in the moonlight, and sometimes with success . . . I suppose I should not boast at law breaking, but protected list or not, if ever I had a chance to pop a rifle bullet into a 'tiger' he would get it, as the saying goes 'right in the neck'.[35]

The human attitude towards this sad and sorry path to extinction is worth exemplifying further, through these three 1937 opinions:

> The Hobart City Council last night adopted a recommendation by the Reserves Committee that £40 should be offered for a Tasmanian tiger (or marsupial wolf) for Beaumaris Zoo . . .

Alderman S. Crisp, opposing the proposal, said that the previous tiger at Beaumaris had fretted itself to death. It was cruelty to take animals from their natural state and cage them up in such a small space . . . When he was told by Alderman Rennie that the previous tiger had lived for 10 years at the Zoo, Alderman L. H. Saunders said that it was 10 years of cruelty to a dumb animal. The man who put it in the cage should have been there himself.[36]

Fancy actually 'protecting' a villainous and repugnant animal of this type. It would be a good thing if the species did become extinct, even at the risk of upsetting the balance of nature.[35]

Sir—I desire to enter an earnest protest against the purchase, by the City Council, of a Tasmanian tiger for the zoo. First, because it will be a further waste of money on a place which has cost and is costing far more than the city can afford; second, because it will mean the keeping in captivity of a frenzied, frantic creature. After the frenzy has died down, it will pace up and down, its whole body expressing the devastating misery it feels, and in the course of time it will die. It will be quite unlike a tiger in its natural state. What possible good can its purchase do anybody? No one will learn anything from it. The idea that people learn natural history from zoos is so out-of-date that an intelligent person would be ashamed to put it forward. I have seen one animal brought in from the bush, and I never want to see another. The frozen despair which its face and its whole body expressed would wring the heart of any person not entirely without imagination. Its frenzy was over, it had refused to eat, but the sight of it will live in my memory to the end of my life. I find it impossible to understand how any man or woman can be associated with such a deed.[37]

11

A BAD FINISH: 7 SEPTEMBER 1936

> Nothing in this whole world will convince me it was not a tiger. I saw the whole of its tiger like shape, its stripes, its well shaped head which I knew I'd never seen before . . . I love dogs and to me it was just another lovely dog I had never seen before. I went to call my husband from bed, but you know men—never interested in anything till after ten a.m., so I have no one to back my story, but I will take an oath it happened.
>
> EDITH THYNNE, TAROONA

In the late 1920s, Labor Party premier J. A. Lyons, a long-serving politician who was later to become Australia's first prime minister from Tasmania, faced certain defeat at the polls. A reactionary, conservative Legislative Council had blocked much of his proposed legislation and to many of his supporters he had betrayed his own ideals, central to which had been his 'powerful attacks on the evils of capitalism and the prospects of war and misery said to be carried within it . . . Lyons had pleased many Tasmanians but disappointed the generation which perceived in the triumph of Lenin and the Bolsheviks a harbinger of social justice and prosperity for all'.[1]

Clearly, Labor's newfangled, radical left-wing sentiments (forged amongst the hardened mine workers of the west and their political hero, the maverick King O'Malley), were a world apart from the equally hard conservatism that had long been practised by the wealthy rural rump of politics. Yet again, through politically laced legislation, the thylacine was to become a victim of the island's Janus-faced nature. Indeed, the animal was unwittingly setting a trend in Tasmanian politics, which has since been characterised above all by bitter battles and stand-offs associated with conservation.

Through the 1920s Clive Lord and others had patiently built up the case for saving the thylacine. The National Park Board had come into being to protect flora and fauna within the park, and then a Tasmanian Advisory Committee re Native Fauna, of which Lord was secretary, was established. The committee made recommendations for federal legislation to wholly protect and to prohibit the export of thylacines. The latter, bizarrely, was in part prompted by the very high prices offered by overseas zoos, against which the Tasmanian zoos couldn't compete. Yet the message still didn't become apparent, neither to the public at large nor to any government, that thylacines were both disturbingly rare and extremely valuable.

This may well have been because the Advisory Committee's recommendations on protection were 'immediately and effectively countered by rural-rump politicians and members of the conservative establishment who prevented, for as long as possible, any positive action being taken to preserve the thylacine, even as irrefutable evidence of the destruction of the species mounted'.[2]

Lord and his colleagues persisted, and in 1928 the new Nationalist Party of Premier John McPhee was presented with further evidence of the animal's endangered status derived from data gathered by 38 police stations across the state. According to Paddle, the government employed a stalling tactic at this point, by having the Advisory Committee recast as the Tasmanian Animals and Birds' Protection Board. This manoeuvre also enabled the government to stack the new Board with its own appointees, 'representing vested interests supporting the timber industry, the hunting and snaring of native game, and the uncontrolled expansion of agricultural enterprise'.[3] As a result, Lord's protection efforts were repeatedly turned down, until at last in 1930 he managed to steer through the partial protection measure banning hunting in December. It also enabled prohibition of exports, which could not be legally stopped while the animal was wholly unprotected.

At the practical level it is clear that this measure in no way helped the thylacine. Reference has already been made to the snarers' general disregard of legislation ordering them what not to catch. Also, the grim years of the Great Depression from 1929 to 1933, perhaps even to the outbreak of World War II, only increased the value of thylacines: as zoo stock, for their skins or, on the sly, between hunter and sheep owner as continued bounty.

In retrospect it is all the more tragic that so few Tasmanians genuinely cared for their unique marsupial predator. They had been proudly exhibited at both the Hobart and Launceston zoos for many years; countless thousands had paid to enjoy viewing them alongside other native and imported wildlife—yet few cared about them enough to become involved in their preservation. The story of Hobart's two zoos, for example, is a story of lost opportunity, not least because of the energy and commitment put into them by a dedicated few, two women in particular. Many thylacines were handled by the zoos. How easily they would have formed a captive breeding stock if they had been

A captive thylacine has been fed a chicken. The absence of visible enclosure wire in the photograph renders the image as an attack upon a fowl house. (Don Stephens)

released for that purpose on Maria Island off the east coast, which was eventually mooted as a sanctuary.

Back in 1895, wealthy Hobart socialite Mary Roberts had, as a hobby, started a menagerie that grew into a successful private zoo, in the large grounds of her mansion Beaumaris in the suburb of Sandy Bay. Her wealth enabled her to buy exotic animals and birds and she became a formidable entrepreneur in the animal trade, while remaining an animal lover—to the extent of forming the Anti-Plumage League in 1910, which campaigned against the killing of peacocks and the like for human adornment.

Her Beaumaris Zoo contained a typical range of specimens—lions, zebras, antelope, polar bears, an elephant, many birds—as well as native fauna, of which the thylacine was a star attraction, as this recollection attests:

> When a child I was taken to see a tiger in the Beaumaris Zoo in Hobart. A fawnish-yellow creature with dark stripes across its powerful hind-quarters and a strong rigid tail, like an Indian tiger but with the head of a wolf and a strangely hinged jaw opening at right-angles, measuring about four feet from the nose to the base of the tail, it prowled with a stealthy lumbering gait round a small netted yard.[4]

As well as displaying thylacines for the people of Hobart (and, it seems, most visiting dignitaries), Mary Roberts hand-reared a number, although it is unlikely that she tried to or actually bred them, since her diaries make no mention of it. She regularly used her zoo to promote animal welfare through education and fundraising. Thus an event for the Royal Society for the Prevention of Cruelty to Animals was attended by the Governor Sir Harry Barron who,

> declaring the function open, said how much the people were indebted to Mrs Roberts for helping the funds of so many institutions by throwing open the zoo on numerous occasions. He had been told that in Tasmania there was no animal that could not be tamed. Mrs Roberts had tamed even tigers, but he would not care to go in with them like she did![5]

In the light of such apparent public feelings for the animal, it is all the more surprising that the later efforts of individuals like Clive Lord would amount to so little.

At her golden wedding anniversary in 1913 Mary Roberts was presented with a solid gold thylacine brooch. That may have been a canny gift, for she did a good trade in them and their carrion-eating cousins the Tasmanian devil—of which she famously remarked, about those favourites of hers, that she was 'possessed by seven devils'. In the north, James Harrison was the island's other main dealer in live thylacines. Mary Roberts was generally able to onsell thylacines for double the prices she paid to snarers and trappers, with the prices themselves rising all the time. In 1910 she paid about £8 per thylacine; this had risen to £20 by 1919.[6]

Beaumaris Zoo was a well-managed, popular success, run for 26 years by an animal rights campaigner, yet the puzzle is that the obvious plight and value of its rarest inhabitant attracted no official interest, as Clive Lord was discovering.

Upon the death of Mary Roberts in 1921, her daughter donated the zoo to the Hobart City Council, which built new premises for it at the Queen's Domain, a large wooded city reserve on the western bank of the Derwent River. There was evidently considerable pride in this new venture and a major feature article in *The Mercury* sang the praises of its educational and scientific status as being up with the best zoos of the Commonwealth, while it had a layout that was 'a veritable fairyland for the animals'. Real effort was made, including the cutting of a cliff to sculpt 'a wonderfully fine rocky spot, where in comparative freedom the lions will be able to prowl at leisure'. A thylacine features as one of three animals photographed for the piece, the caption being: 'It is intended to keep [thylacines] in the Zoo. They are members of a rapidly diminishing race'. But presumably that was neither here nor there to the newspaper's readers, who probably were better conditioned by the times to accept a decidedly warped reality:

> To [the Curator's] sympathetic treatment, the devils, usually the personification of ferocity, the eagles, and other animals and birds, have responded wonderfully, and they have come to regard their prison as merely a playground where, free from that ceaseless persecution attendant on their natural conditions, they are fed and cared for . . .[7]

The new zoo's curator, Arthur Reid, was highly experienced and diligent and his daughter Alison worked closely with him, taking on the role of de facto curator after her father's death. She worked tirelessly; to the public she became a well-known figure at the zoo with her pet leopard. She also was the taxidermist of the Tasmanian Museum and Art Gallery, of which Clive Lord was director. Unfortunately, like Lord, her efforts to promote the welfare of animals, in particular the zoo's thylacines, were to be stymied.

Meanwhile, the puzzle of a public that admired thylacines while being indifferent to their fate is nowhere better illustrated than in a piece of journalism that reads like a cheerful obituary. It appeared in the 70th anniversary edition of *The Mercury* on 5 July 1924, in a special celebratory supplement entitled 'Preserving Our Native Animals and Birds: Phenomenal Success of the Beaumaris Zoo'—with the thylacine taking pride of place at the head of the article, followed by a summary of the history of zoos and of Beaumaris Zoo. It is a cruel celebration:

> 'Living Fossil'. How would you like to be called a living fossil? Not very complimentary, is it? That's what some biologists (people who study animal life) have called our tiger, or wolf. You see, the tiger, as we shall call him, is a very old beast—much older than you would imagine. You have only to look at him to see that: His legs are relatively short, and body somewhat elongated, head narrow, with little brain capacity, a bad finish about the hindquarters; in fact, an unproportioned experiment of nature quite unfitted to take its place in competition with the more highly-developed forms of animal life in the world today. We have evidence that his ancestors enjoyed a distribution covering a considerable portion of the globe . . . European scientists found fossils of the creature, and

> wondered what he was like, when, behold, down in a little island round the other side of the world he was found to be alive, eking out a precarious existence, it is true, but spending his last days on a globe which he had inhabited for so long, in comparative peace . . .[8]

Lord was still patiently trying to persuade the Animals and Birds' Protection Board to recommend total protection when he died in 1933. Such had been his dedication and persistence that the Board was spurred to rethink its views, and a new committee was established to look into the issue of saving the species. At the same time, barely a kilometre away at the city's edge on the Queen's Domain, Alison Reid was fighting a losing battle to save just one thylacine, apparently the very last in captivity, whose story typifies the species' dreadful stroke of ill-fortune in having had Europeans invade its island sanctuary.

According to legendary thylacine devotee Col Bailey:

> It is the 1924 capture of a female tiger and her three cubs in the Florentine Valley that entered Walter Mullins into the history books. First snaring the mother in a foot snare, Mullins had only to patiently wait around for her brood to come looking for mum before taking captive the whole family. It was to prove a most notable capture, and not even Mullins could have foreseen the consequences. He somehow managed to get them to the Fitzgerald railhead from where almost all the tigers caught in the south-west finished up after 1917. From there the group were transported to the Beaumaris Zoo . . . A deal was soon struck with the Curator, Mr A. Reid, and the four thylacines changed hands for the then princely sum of fifty-five pounds. Mr Mullins no doubt left the tigers to their fate little realising that he had just sold what was to become the last captive tiger, and indeed as some would tell us, the last Tasmanian tiger on planet Earth.[9]

Either this young thylacine, caught by Elias Churchill in 1933, is 'Benjamin', the erroneously named alleged last-known living member of the species ...

Or this is 'Benjamin', caught as a cub by Walter Mullins in 1924 and surviving twelve years in Beaumaris Zoo. Such longevity is highly unlikely. And Robert Paddle's research suggests this animal died in 1928. (Above: Robert Paddle, *The Last Tasmanian Tiger*, page 199; below: Collection Tasmanian Museum and Art Gallery)

Guiler's records also indicate that this alleged last captive thylacine was purchased by Beaumaris Zoo in February 1924 and died at the age of twelve years and seven months, according to the authority of Hobart City Council Records.[10] However, just to add to the surfeit of controversy surrounding the creature, and the last one in particular, popular opinion suggests that it was caught by Elias Churchill in the Florentine Valley in 1933. Paddle quotes both Alison Reid and the highly respected naturalist Michael Sharland, who was also associated with Beaumaris, in support of this date, and avers that the film footage, 'together with other still photographs, clearly identifies the last specimen as a mature but relatively young adult female'.[11]

The 1981 Beresford and Bailey book had a presumably unintentional bet both ways on the issue: 'By the early 1930s, only one tiger of the collection remained'.[12]

As if this was not enough to lob on the apparent last-known of a species, Paddle also uncovered a seeming act of deliberate misinformation against that particular animal many years later, in which a man called Frank Darby falsely claimed to have been its last carer and that its name was Benjamin, a name which has entered Tasmanian folklore.

The puzzle of the last-known thylacine had one final twist, uncovered nearly 90 years later. As reported in the December 2022 Tasmanian Museum and Art Gallery newsletter:

> Ground-breaking new research has solved one of Tasmania's most enduring zoological mysteries: what happened to the remains of the last known thylacine, or Tasmanian tiger? . . .
>
> Researchers Robert Paddle and Kathryn Medlock reveal that the remains came into the collections of the Tasmanian Museum and Art Gallery (TMAG) in 1936 but had until now remained unidentified. The last known thylacine died in the

Beaumaris Zoo on the Queen's Domain, Hobart, during the night of 7 September 1936. However, Dr Paddle, a comparative psychologist from the Australian Catholic University, and Ms Medlock, Honorary Curator of Vertebrate Zoology at TMAG, have discovered that this thylacine was not the animal previously identified as the last thylacine in photographs and film—the much photographed and filmed specimen was in fact the penultimate thylacine.

Dr Paddle said the actual last thylacine, or endling of the species, was an old female animal that had been captured by trapper Elias Churchill from the Florentine Valley and sold to the zoo in the middle of May 1936. 'The sale was not recorded or publicised by the zoo because, at the time, ground-based snaring was illegal and Churchill could have been fined. The thylacine only lived for a few months and, when it died, its body was transferred to TMAG. For years, many museum curators and researchers searched for its remains without success, as no thylacine material dating from 1936 had been recorded in the zoological collection, and so it was assumed its body had been discarded.'

Ms Medlock said the last thylacine's arrival at TMAG was verified by the discovery of an unpublished museum taxidermist's report dated 1936/37 that mentioned a thylacine among the list of specimens worked on during the year. This led to a review of all the thylacine skins and skeletons in the TMAG collection, and the subsequent discovery of the last animal. 'The thylacine body had been skinned, and the disarticulated skeleton was positioned on a series of five cards to be included in the newly formed education collection overseen by museum science teacher Mr A. W. G. Powell,' Ms Medlock said. 'The arrangement of the skeleton on the cards allowed museum teachers to explain thylacine anatomy to students. The skin was

> carefully tanned as a flat skin by the museum's taxidermist, William Cunningham, which meant it could be easily transported and used as a demonstration specimen for school classes learning about Tasmanian marsupials.'
>
> TMAG Director Mary Mulcahy said the last thylacine's tanned flat skin and disarticulated skeleton, still attached to the five cards created for the education collection, were now on display in the museum's thylacine gallery. 'It is bittersweet that the mystery surrounding the remains of the last thylacine has been solved, and that it has been discovered to be part of TMAG's collection. Our thylacine collection at TMAG is very precious and is held in high regard by researchers, with the museum regularly receiving requests to access our mounted specimens as well as thylacine bones, skins and preserved pouch young. Our thylacine gallery is incredibly popular with visitors and we invite everyone to TMAG to see the remains of the last thylacine, finally on show for all to see.'[13]

Beaumaris Zoo suffered the effects of the Great Depression. Paid staff were laid off and replaced by work-for-the-dole 'sussos' (sustenance workers), with limited understanding of and interest in the animals. Its gate takings dropped. Alison Reid herself received short shrift, possibly because she was a woman: unpaid, threatened with eviction from her cottage at the zoo, and refused possession of a key to gain access to care for the animals after hours. It seems that despite her best efforts, including a special plea to the Hobart City Council on behalf of the zoo's last surviving thylacine, the animals were increasingly neglected.

There could only be one outcome:

> With genuine distress in her voice, Alison recalled to me the last weeks of her life at the zoo in 1936. Powerless, keyless and shortly to be dismissed from the zoo and turned out of her

> home, she listened at night to the distress calls of the zoo's remaining carnivores: the last thylacine, a Bengal tiger and a pair of lions, all too frequently locked outside in the open to face the cold, rain and snow of the Hobart winter.[14]

There exists 62 precious seconds of film footage of the penultimate thylacine, taken by David Fleay (who was bitten on the buttock while filming). Just two months earlier, legislation had passed through the Tasmanian Parliament declaring the species to be wholly protected. That was not and is not the final irony in the thylacine story, not by a long measure—but as a practical demonstration of human folly masquerading as progress, it surely has few peers. Did the politicians in 1936 give any thought to their elected forebears of half a century earlier? Guiler did: 'Those twelve parliamentarians who voted for the Bill to pay a bounty without enquiring into the authenticity of the facts and figures given have a great deal to answer for in the history of conservation in Tasmania'.[15]

Tasmanian east coast resident, artist Vita Brown, 86, has a very clear memory of visiting Beaumaris Zoo and seeing the last known thylacine. In her opinion it was very old. She further describes it as being emaciated, lonely and, so far as this is possible with an animal, wearing 'a look of despair' as it paced up and down. The sight of it upset her so much that she had to leave the zoo and never went back.[16]

12

A LOST OBJECT OF AWE

> In about 1922 or 1923 I rode up the Lake Road to see how deep the snow was . . . It began to snow again so I turned my horse for home. When I got just level with the cave my horse stopped at the sound of a growl & there on my right was a full grown Tiger showing its teeth and still growling. It was 2 to 3 feet from me, & the passage between the rocks into the cave was too narrow for the Tiger to turn around. My horse blew its nostrils & the Tiger backed himself back. My horse jumped forward. Tigers are so shy & cunning . . . I was very lucky to have been so close to a Tiger.
>
> Mrs Linda O'Shea, Launceston

'Without an end to clearfelling, our old growth forests will share the fate of the Tasmanian tiger: a lost object of awe, one more symbol of our feckless ignorance and stupidity.' These words were delivered by internationally-acclaimed Tasmanian novelist Richard Flanagan to a 4000-strong protest rally in Hobart in August 2001, staged outside the casino-hotel hosting the Australian Labor Party state conference. Other notable speakers included Greens Senator Bob Brown and actor Rachel Griffiths. The message was clear and loud: directed at the incumbent Tasmanian Labor government, it demanded cessation of the practice of converting the island's ancient forests into Japanese toilet paper and like products, by way of woodchipping.

Conservation has long been the touchpaper of Tasmanian politics. It traces back to the political demonisation of the thylacine in the 1880s. The animal was neither pest nor vermin but was singled out of its environment for the apparent economic betterment of the few, by a conservative mindset demanding the continued conquest of wilderness. That Tasmanians were unable to save it from extinction—that they did not even try until well into the twentieth century—eventually spread an enormous pall of guilt, regret and sorrow. The thylacine's demise left a terrible legacy—perversely, not unlike the so-called convict stain which many families tried to hide, also until well into the twentieth century.

In the end, the tragedy of the thylacine created a rare energy in the island, a collective determination to henceforth resist environmental damage, degradation and obliteration. This has caused much conflict, conflict which still flares and will continue to do so. In some of Tasmania's environmental friends and defenders the thylacine engenders high emotion, in others reasoned logic, but both start from the same premise—that the animal was maliciously hunted until close to extinction.

Important though it is, the dramatic touch of having a single date to signify the 'death' of a species obscures the facts of the thylacine situation as it then applied. Tasmanians did not go into collective mourning on 8 September 1936. They had been ignoring and persecuting the animal for so long that thoughts of its permanent loss occupied but a few minds. Nor, in fact, had the animal become extinct. But those who were concerned at least now had that date as something tangible to mount a case for action. The last known captive thylacine was dead. An attempt had to be made to disprove the alarming possibility that it might, indeed, have been rendered extinct.

To that end the Fauna (Animals and Birds' Protection) Board sanctioned two searches in 1937. The first scoured the Middlesex Plains and adjacent areas and the second went deeper into the western mountains, in the rugged Frenchman's Cap area. Both of these searches found prints, and some sightings were also investigated—but that was all. It set a depressing precedent for the many searches and expeditions, official and private, that were to follow well into the twenty-first century.

Still, there remained a fair amount of optimism that remnant groups of thylacines ought to be alive and well in the more remote, untouched and inaccessible parts of the island. Undisturbed, they would hopefully rebuild a sustainable population. But rebuilding human affairs in the wake of the Great Depression came first. Water, wood and minerals were the island's principal natural resources; they needed to be yoked ever more firmly to the economy, to create much-needed jobs and revitalise the export market. In the years preceding World War II a major hydro-electric power station at Tarraleah and newsprint mills at Boyer and Burnie were built. From now on, getting the wood and harnessing the water meant entering those remote and inaccessible parts. The heavy machinery developed during the war would later make this all the more possible. Thus, the battlefields of the developers and conservationists were waiting, with the thylacine's last refuges and its very future, if indeed it still had one, integral to the conflict.

History shows that the first true Tasmanian conservation campaign began in the one place where there ought to have been no need for a campaign—the Russell Falls–Mount Field National Park, where in 1949 the government of the day overrode existing protective legislation in order to permit

7000 acres (2000 hectares) of its Florentine Forest to be logged for the newsprint company ANM. It was a surreptitious move, but it caused a public outcry and led to the formation of the state's first conservation group. That group did not achieve its major objective but laid the foundations for future public action against government and business acting together to put profit before environmental considerations—of which the bounty scheme had been but the first example.

At the end of World War II the director of the animal sanctuary at Healesville in Victoria, David Fleay (he who shot the famous thylacine footage), was given official Fauna Board support to head an expedition to capture a thylacine pair for breeding purposes. The group covered a significant amount of territory in the west, their baits including bacon, live chickens and meat. Again there were prints, sightings obtained from local bushmen, and one evening the party listened to a thylacine calling. Fleay's major recommendation after the expedition was that snaring and trapping be prohibited across the south-west. He felt also that poisoned devil baits were surely being taken by thylacines. It was another alarm bell, from an experienced zoologist.

Yet the official position on thylacine protection seemed to be somewhat muddled, even counter-productive, as a 1946 letter, written by the director of the Queen Victoria Museum in Launceston to a snarer, suggests:

> Dear Sir,
>
> I have received your letter about the Tasmanian Tiger which you have heard near your house.
>
> The Tiger is a protected animal and can only be killed or captured by special permission. This Museum holds such a permit and will authorise you to take the animal on its behalf

on the understanding that if successful you present the living animal or its carcase to this Museum. Would you please let me know if you are agreeable to this; when exhibited it would be shown as a gift from you.

Your local police trooper would be advised that you were trapping for the Tiger on our behalf.

You could try to trap the Tiger in the following ways:-

(a) Build a large cage with an entrance passage, screen it with bushes. Drag a trail leading to it with bait; you can use the intestines of a sheep or use rabbits and so on. Put the bait in the cage after you have dragged the trail. Set a trip catch at the end of the passage which would release a door set at the entrance to the passage when the Tiger knocked it.

This trap would probably be the most satisfactory but would involve considerable work.

(b) The second method is to drag the bait along a run and set strong springer snares on this track. The snares should be of the type which will catch the feet, not a strangling snare, as we want to get the animal alive.

If you catch the Tiger let the Police Trooper or myself know *at once*.

Please let me know if you will be trapping. If there is anything you are not sure about tell me when you write.[1]

The classic and frequently reprinted 1941 work of Ellis Troughton, *Furred Animals of Australia*, has already been referred to; his recommendation for the species during that early period of uncertainty was also, not surprisingly, equivocal: 'Though the interests of the settlers must be the first consideration, there is urgent necessity for an adequately controlled reservation in some remote fastness to ensure the survival of such a unique animal.'[2] Troughton also kept alive the questionable marsupial trademark:

'Keen observers have stated that, if hard pressed, it will rise on the hind-legs and hop like a kangaroo. This action may be for clearing obstacles in haste, but the appearance of the hind-quarters supports the idea'.[3]

Alas, there were those who continued to wreak havoc upon it, if only in print:

> Zoologists are fascinated by the Devil and Wolf-Tiger because these are the only marsupials of their kind left in the world . . . For the average visitor, however, the pair have only a museum interest: they are never seen by the tourist, although a Devil may be found in a zoo. The man who brings a Tasmanian Wolf to an American zoo will make a fortune. Speaking of them from the purely personal standpoint, both animals are stinking and vicious, and Tasmanians are entirely agreed that outside zoos they are not missed from the Tasmanian landscape.[4]

By the early 1950s Tasmania's population had topped the 300 000 mark, thanks in part to immigration—mostly English and Dutch—and the economy had benefited from postwar economic booms, particularly in farming. Modernisation through mechanisation greatly increased agricultural output, which had knock-on effects:

> Transport and communications also changed radically . . . Roads partly based on ancient Aboriginal tracks and the easiest path for horse or bullock teams, were straightened, old settlements bypassed and modern highways pushed through to facilitate the speed and easy travel of cars and semi-trailers. In the bush, chain-saws and caterpillar-track vehicles took over from axes, horses and sheer manpower.[5]

None of this could possibly bode well for a perilously endangered species reliant for its survival on a total absence of human interference. But interest in it continued, to the extent that at the official level there were fears that thylacines were, more than ever, being pursued for personal gain. Was it turning into a legend? The worried Fauna Board 'became more and more sensitive to people operating without [its] knowledge or approval. By the mid 1950s, it determined to control all tiger searches and started its own investigations'.[6] This was a logical and welcome advancement as a protective measure. But, in a fact-is-stranger-than-fiction move, the Board was once again undermined by its own government, which secretly negotiated with a mainland media organisation and *Life* magazine to sell the visual rights arising out of any capture. Effectively, the government was auctioning off copyright in the thylacine. Guiler, the then chairman of the Board, states that the Board was 'greatly distressed and annoyed' and exerted pressure on Labor premier Eric Reece (who was later to feature in more tiger business), forcing him to pull the plug on the deal.

So it was that between 1957 and 1966 there were eight official searches, investigations and expeditions, with Guiler involved in all of them. While the localities and methods differed, what they had in common was justified optimism at the outset and negative results at the end. The first search was prompted by apparently unmistakable thylacine behaviour in the Derwent Valley, involving sheep kills, with one lamb the victim of precise entry through the rib cage to remove the liver.

A 1958 sighting in the Ben Lomond area, with a confirmed thylacine print, led to a large baited trap being set up by Guiler and a team near a group of cave lairs. Nearby trees had sheeps' heads nailed to them and at a later stage bullock livers were dragged around. Again, all to no avail. Scats collected in the area were analysed and judged to belong to Tasmanian devils.

Like so many Tasmanian 'localities' (defined as having 200 inhabitants or less, with some facilities), Trowutta is little known and off the beaten track. South of Woolnorth, it is an entry point for the Milkshake Hills Forest Reserve and the uninhabited wilderness beyond and is very much thylacine country. Here Guiler investigated an alleged sighting by an experienced snarer who claimed to have seen a thylacine on, of all things, the woodheap at Trowutta Mill. Again, the investigation went nowhere, but later in the same area a large unknown animal escaped from a snare set by James Malley (one of the state's major tiger resurrectionists). Guiler's opinion was that Malley's snare had not caught a thylacine because the disturbance at the scene didn't fit with 'the agreed opinion of thylacine hunters that they just gave up in a snare'.[7] Such is the frustrating mystique of the animal that this contrasts directly with the experience decades

Eric Guiler with a small stuffed thylacine head. Guiler devoted much of his adult working life to the thylacine, trying to protect it through habitat preservation, searching for it and writing about it. (The Mercury)

earlier of an experienced bushman who bought and sold thylacines to zoos: 'Mr Bart thinks he caught one himself in a snare, judging by the way it had chewed its way out, making a terrific mess of the surrounding vegetation. Dog very scared'.[8]

In the late 1950s and early 1960s Guiler led a number of expeditions and investigations at Woolnorth. This locality held promise for a number of reasons, not least its being largely unchanged since its initial use for sheep pasturage in the 1830s. These searches were notable for the first use of fixed cameras, purchased for the purpose from RAAF disposal stores. At this time too, a little further south at Sandy Cape, there occurred a famous and controversial incident in which two fishermen were awoken by a sound and one of them, seeing a pair of eyes glowing in the dark, struck out with a piece of wood. The next day they discovered that they had killed a thylacine. Had they? Whatever the truth, from that point their story became the bush equivalent of an urban legend, in which they stashed the carcass underneath a sheet of roofing iron and went off to tell of the deed. Alas, when they returned it had been stolen and was later, apparently, offered for sale to Sir Edward Hallstrom of the Taronga Park Trust, who refused it, after which it was dumped in the sea. (Hallstrom had earlier conducted three unofficial thylacine searches and it was these that prompted the Fauna Board to step in and mount searches of its own.)

Guiler made two more west coast trips, the first funded by the government, with Premier Reece hoping for good thylacine publicity. Thousands of leg snares were fruitlessly laid. The second—supported by a World Wildlife Fund grant—was inspired by the government's acquisition of Maria Island as a sanctuary where captured thylacines could be released. This happened in 1966, after considerable lobbying by Guiler and his Board. In turn, the Board was itself being vigorously lobbied at this time by a public group seeking protection for a much bigger area, the south-west wilderness. That group, the South-West Committee, had recently been formed in response to growing public and commercial interest in one of the world's most unspoilt tracts: the great swathe of mountains, forests and rivers stretching from the central plateau to the remote south-west coast.

The Board was under pressure because it stood between the South-West Committee and the Hydro-Electric Commission (HEC), which had for a decade been investigating the area's great commercial potential. Already there existed, at the very core of the south-west, the Lake Pedder National Park (legislated in 1955 thanks to the efforts of the Hobart Walking Club). What might come next? The committee wanted a sanctuary declared, but that would have shut out commercial exploitation. The three parties compromised and agreed upon a 646 000 hectare Faunal District encompassing Pedder:

> Dr. Guiler said Tasmania once again had gone to the front in conservation programmes. With the exception of the Northern Territory, it would be the biggest area reserved for wildlife in Australia . . . No one will be allowed to take any animal or bird from the area, or enter it with guns, dogs, traps or snares without the written permission of the Board.[9]

It may have seemed like a comprehensive victory in the belated attempt to save the thylacine, but in reality this deal marked the first—and decidedly least bitter—three-way contact between Tasmania's officialdom, its economic oligarchies and its conservationists. The fight for Tasmania's future was about to start, and at its heart was Lake Pedder. Extinction, exploitation, preservation, had inevitably come together. As was so wryly reported in *The Mercury*:

> People fighting to preserve the South-West as a wilderness reserve were grabbing the Tasmanian tiger by the tail, the Premier (Mr. Reece) said yesterday . . . He said the south-west had 'a few badgers, kangaroos and wallabies, and some wild flowers that can be seen anywhere'. An Opposition member interjected: 'And the Tasmanian tiger'. Mr. Reece: 'We haven't been able to catch up with him yet. These people (preservationists) are grabbing him by the tail when they grab this issue.'[10]

Reece had handed the Fauna Board £2000 to find a thylacine; his government could happily give a little on one side, knowing how much would be coming in from the other.

In the nineteenth century the thylacine had been deemed a negative, dangerous part of the natural environment. But when in the twentieth century that environment came to be considered a highly lucrative and therefore positive resource, it was the people trying to protect it who were now considered negative and dangerous. Tasmania's battles over water and wood, in the ever-lengthening shadow of the tiger, were set to begin in earnest. The proto-conservation groups were under no illusions as to what they were up against. Regarding water: 'Hydro-industrialisation is the policy of building dams to provide electricity for heavy, energy-intensive industry. It is a policy that pleases many. Unions get construction jobs, the Hydro-Electric Commission gets a bigger empire, the companies get heavily subsidised power and politicans get re-elected'.[11] Regarding wood:

> A single tree, 70 metres tall, contains almost 100 tonnes of usable timber. Carefully graded and converted this will make: enough thinly sliced decorative veneers to panel the walls of a four-storey hotel, *plus* enough sawn timber for the framing and roof trusses of an average family house, *plus* enough solid wood to make a set of

> household furniture: tables, chairs, beds and cupboards, *plus* after all that, enough pulpwood to photocopy the complete works of Shakespeare over 3,000 times.[12]

Much has been written about the valiant but vain effort to save the original, unique Lake Pedder from being drowned in late 1972–early1973 to create a hugely enlarged body of water for the HEC's Gordon Dam complex power scheme. Like the thylacine, the lake with its pink-white quartzite sand beaches has iconic status, yet it too no longer exists—although, in a somewhat spooky echo of thylacine sightings, its original shape can sometimes be seen from the air (and some vow to restore it one day). But it remains the other vivid Tasmanian symbol of the power of vested economic interests over environmental considerations.

Guiler noted that the flooding of Pedder destroyed 'useful' thylacine habitat; over and above that, the controversy destroyed a Labor government and, in 1972, created the United Tasmania Group, the world's first green political party. Indeed, politics and the environment had become fused 'The struggle to save Lake Pedder did a great deal towards the awakening of Tasmanians to their natural heritage. Thousands who deeply regretted the destruction regretted also the political and administrative machinery that permitted it despite such enormous opposition'.[13]

Who was to blame? The answer to that question dates back to the first European settlement-invasion of Van Diemen's Land, since when the island has generally been at the mercy of ingenuity in the name of progress. Thylacines were a great target and politicians duly took aim, and succeeded, when in 1887 that narrow vote was cast to eliminate them. Nearly a hundred years later, politicians had it in their power to act nobly—alternatives to the flooding of Lake Pedder were tabled and viable—but, in another echo, a bad one harking back to those reactionary east coast grazier kings, the stayer Eric Reece and his old guard 'were simply bewildered by the growing wave of public feeling for the natural environment. Pedder foundered largely because of the pride of old men who had been left behind by changing times'.[14] Paradoxically, a lake had been drowned.

Meanwhile the search for the thylacine went on. In 1968 the biggest expedition to date was mounted, by James Malley—then a Smithton real estate agent—and Jeremy Griffith, a Sydney zoologist. Over a four-year period, throughout the heat and emotion of the Pedder campaign, they worked assiduously towards proof of the animal's continuing existence, sometimes aided and sometimes rebuffed at the official level, and always struggling to find funds to keep the search alive. In 1972 they were joined for a period of nine months by a young Sydney medical graduate, Dr Robert Brown, who ran their Launceston-based information centre. The three were briefly known as the Thylacine Expedition Research Team. When that team disbanded, Brown wrote: 'The enormity of the thylacine's importance in our natural history can be seen as perhaps even greater than that of the odds of there being any members of the species left in Tasmania'.[15] It was a pessimistic summation of yet another failed attempt, but with a new focus. Even as Pedder was drowning, the thylacine was emerging as a beacon, its 'enormity' in the Tasmanian story sheeted home to that same Bob Brown, who then settled in Tasmania and created another kind of history, becoming the world's most influential Greens politician. The Tasmanian Wilderness Society, which Brown co-founded as a result of Pedder, determined that the mistakes of the past must not be repeated. In Brown, the Society had a formidable and forthright voice. He and other protection advocates wanted only preservation. As ever, they battled profit: many decades earlier at the turn of the twentieth century the thylacine had officially been deemed the world's most expensive animal, valued at 75 guineas a pair—the runner-up, the fabulous Bengal tiger, a mere 60 guineas a pair.[16]

13

WE WAKE UP TOO LATE

> It was now 1946 . . . this man Mark McCloud greeted us with the words 'Welcome sportsmen, you are now in Balfour'. Mark was nearing 80 years of age, his clothes were patches upon patches, a shaggy white beard and long hair covered most of his face, but his tall figure now bent with age told of a once hard strong man. We stopped the night with Mark, and that in itself is a story but to hasten on . . . It must have been near midnight when I went outside before turning in for the night. The night was still and sounds carried far, when suddenly a sound I had never heard before came from a nearby hill. The sound was clear, not a bark but rather a high pitched yelp. Immediately I called to Mark and Ray and Mark had no hesitation in stating that it was only a tiger after the hindquarters of a roo he had left hanging in a tree.
>
> C. L. Smith, Tarraleah

In the latter part of the twentieth century, interest in the thylacine continued to grow. Sir Edmund Hillary, in Tasmania to climb its wild peaks, expressed a hope of seeing one. Animal welfare celebrity Brigitte Bardot wanted to search for it but funding problems prevented her. However, in 1978 the newly established Australian chapter of the World Wildlife Fund put $55 000 into a proposed search, to be run jointly by Eric Guiler and Steven Smith of Tasmania's National Parks and Wildlife Service, formerly the Fauna (Animals and Birds' Protection) Board.

The two men worked assiduously over the next two years. Much of the funding was spent on sophisticated automatic cameras which were strategically placed in a wide variety of likely locations. Alas, while virtually every other species of Tasmanian large mammal was filmed, not one thylacine showed up. If proof was required that they were no more, surely it had come through the patient lenses of those expertly positioned cameras. Smith published his report in 1980, declaring the species extinct. Yet very many Tasmanians continued to have a tiger story confirming its presence.

International interest in the thylacine may be gauged by this lovely if inaccurate 1974 educational snippet, in which a North Carolina, US, newspaper invited students to ask natural history questions, in return for which the selected student received a 'complete 20-volume set of the *Merit Students Encyclopedia*'. Kathleen Sevy, age 13, of Clearfield, Utah, asked: 'What is a Tasmanian Wolf?'

The columnist, 'Ask Andy', explained:

> A century ago the settlers of Tasmania called him a hyena, a zebra wolf or a tiger wolf—and hunted him almost to extinction. As his numbers decreased, so did his bad reputation for making off with domesticated sheep and poultry. Nowadays, the rare survivor is hunted to make sure he survives . . . The thylacine is three feet long, plus an 18-inch tail . . . Sometimes he hunts alone, sometimes with a relative. The partners communicate with doggy growls and barks and sometimes the thylacine whines like a puppy . . . Though rare, the thylacine no longer is listed as an endangered species.[1]

In 1981 the first book devoted exclusively to the thylacine was published—although, at a mere fifty-four pages, and concentrating on searches to that point, it was hardly intended to be comprehensive. Written by Canberra-based historian Quentin Beresford, and Garry Bailey, then a journalist with *The Mercury*, its blurb posed the inevitable question: 'Does the Tasmanian tiger still exist?'[2]

This little book, important in its own right, gave rise to an equally important historical record of the thylacine. To publicise and celebrate its release, publisher Dan Sprod organised a Tasmanian Tiger Competition in September 1981, in conjunction with the now-defunct *Tasmanian Mail* and a hotel chain. The newspaper advertised it as 'a contest for the best true story relating to the Tasmanian tiger', a point emphasised in the conditions of entry: 'The content of the story must be fact not fiction'. Clearly, the exaggerated yarns known to be out there were not required. A limit of 500 words was set, with these further terms:

> Stories which qualify for entry include personal recollections of the writer (or friends or relatives), stories passed down from those no longer living, or accounts told in letters or diaries from times now past . . . The inclusion of a factual background, such as dates, and the names of persons and places, is recommended . . . Interesting facts about this unique animal are sought. Skill in writing is not needed to win the competition . . .

The contest attracted 39 entries: not many, perhaps, but they are the invaluable experiences of Tasmanians talking about their special, tragic, marsupial predator. Diligent, honest and full of character, they are heartfelt and add to the richness of the island's story.

First prize was two nights' accommodation at a Four Seasons Motor Inn in Tasmania and a copy of Beresford and Bailey's book;

five runners up also received the book. It was with significant foresight that Sprod decided that 'At the conclusion of the contest all stories entered will be deposited in the State Library [Archives Office] of Tasmania for use in future historical research'.[3] Many stories are handwritten, by elderly entrants; legibility is occasionally a factor in their reproduction. Those that appear here have been transcribed exactly as they were written.

From Betty. Holmes, East Wynyard

Every time I hear the Tas Tiger mentioned I am filled with remorse when I think back of a thing my mother and I did. As early as I can remember my collector Uncle lived with us. Occupying the largest bedroom for his displays, of mineral, & marine, and the 'Tiger skin'. It held pride of place in front of the fireplace. He loved to take visitors to his room to see his various collections. How my mother hated that skin, she was really ashamed of it . . .

At my Uncle's passing in 1956, that treasure was first on the fire, as my mother got rid of his rubbish, as she put it. When my Uncle became ill, an old friend he had not seen for years called to see him, & was Talking about the skin, & I can remember his saying to my Uncle, My word Alf, you'd do well to catch one today, I reckon they have nearly been 'wiped out'—How pitifully true those words have become—I can see it lying there on the floor as tho it were yesterday, sort of fawn colour as I remember, and its black stripes—across the back & measuring about 4 [feet] long give or take a little. and hating it because my mother did, and so destroying a museum piece for the future generations.

From Raymond Melville Lawes, Burnie

. . . This whole area is known as Cuckoo Valley where we lived (a family consisting of Mum & Dad and seven kids.) My father

Alfred Lawes cleared and used a lot of bush land under these hills. Being a very poor family my father and us older boys done a lot of hunting during open season—often snaring well into the mountain country and it was up there about 1930 when I was fourteen years old that I saw my first Tiger from about six feet.

I had climbed onto a big log one foggy afternoon to look at a snare down the lower side—sure enough there was a wallaby but sitting there eating was what I thought was a striped greyhound. It looked at me then took off in a line for the myrtle forest.

Back at camp telling my father he laughed and said—no dog that. That's the damn Tiger. Been eating wallabies up there for some years now. Said I should have shot it. (I had a rifle with me and was a good shot.)

We—my elder brother Lawrence & myself hunted every winter and during the next couple of years we became pretty good bushmen—so much so that we graduated with our uncle Len James to the plain away behind the mountain range. This was known as Maurice Plains and was rough and rugged country. Stretches of myrtle forest wet tea tree flats—the biggest and tallest I have ever seen. In between rising ridges of gum where the possum fed. All this led down to stretches of plain dotted with many old logs. The whole area abounded with wallaby and other game.

It was while camping and hunting here that we had many experiences with our Tiger friends. Uncle Len thought there was a large family around.

Our dogs disappearing back to camp was a common occurrence—we had some pretty large fellows—kill kangaroo and badger easily.

Certain nights somehow seemed to be 'Tiger nights', a strange eerie quietness seemed to come over and uncle would say—Listen—and then the dogs have gone home. We would sit quiet for a while and then light up the spot lights and slowly look all around—fleeting glimpses of glaring green eyes in the distance—left—front right & even behind told the story. They never came very close and we never had any fear of them . . . This was really Tiger country and today nearly fifty years later providing forestry workers haven't moved in too close I see no reason why there could not be some chance of a Tiger still living somewhere in that country . . .

From Kath Doherty, Ulverston

It was a cold winter evening in late June 1916 when my young brother Jack rode through our home gate with his arms clasped around a live 'hyena'. He carried it on the pommel of the saddle, the lower part in a potato bag and its jaws firmly clamped in a mussle. He was just twelve years old and always remembered the thrill of mixed fear and excitement of the taking, the mussling and the trussing that day.

My father, Frank Upston, had set a snare run that terrible winter in the World War I year, to help keep something on the table for five children and another already overdue, who arrived on 9th July. As well as the ready cash for the skins, we often had and liked wallaby tail soup and patties made of mince hind leg & bacon.

This snare line ran through the area now called Meunna while we lived in Lapoinya, where the forestry area building now stands.

He knew the 'Hyena' was on his scent each evening. As he left the scrub he could see the movement just a chain behind. It never approached any closer and always stopped (or the movement of the low shrubs stopped) when he stopped.

He told us of this for several days he had been offered £25.00 for one alive by James Harrison and decided to take it

This juvenile, restrained by a collar, became a Hobart Zoo inmate. It appears to have 'given up', as some trappers said happened when captured. (Collection Tasmanian Museum and Art Gallery)

alive. He made and set strong snares in places he knew it to frequent. He also fashioned and made a rawhide mussle, which he carried each day.

When he did take it, a half grown male, it was not in the 'hyena' snare, but a light one on a frail springer, in a place where he did not expect it roamed. My young brother had ridden into the scrub to meet dad and carry out the skins, so together they trussed it and brought it home. We had it on display till next mid afternoon in our house and I think all the neighbours for miles around came to see it. It did not take any food, but we sent live joey wallabies for it, on its trip to Sydney, where it went to the Sydney Zoo. Most potatoes from the Wynyard–Table Cape area went to the Sydney market by a small fleet of Auxiliary Sailing Vessels from the Wynyard wharf in those days, and, our 'tiger' was shipped from there.

From Gordon Graham, Swansea

It was just after the turn of the last century that the late Mr William Cotton came into the town of Swansea leading a Tasmanian Tiger, most people at the time were scared of the animal, and were amazed to see a person doing such a thing.

At the time Mr Cotton was living on a small property at Swansea named 'The Hermitage', it was winter time, and the taking of kangaroo, and wallaby was permissable. So it was that trapping, and snaring was in full swing throughout the district.

William Cotton that winter had snares set about 4 miles west of Swansea at a place called 'The Sugar Loaf', and on visiting these snares one morning found he had caught a tiger.

The animal appeared rather docile, and showed no signs of fight, so it was then that Mr Cotton thought of the idea of trying to get the animal back to Swansea alive so that he could show of[f] his prize.

After some consideration he cut a short pole about five feet long, and to the end attached a piece of rope from his carr[y]ing bag, and with a noose made on the end slipped it over the tigers neck, held him at bay, cut the snare, and set of[f] to Swansea leading the tiger with him.

He had great trouble to get the animal to travel, but after going a few hundred yards the animal started to act just like as if it was a dog, and followed along beside him for the rest of the way to Swansea with the least of trouble.

Back at 'The Hermitage' the tiger was tied in the stall of a horse stables with a collar, and chain, where it was fed, and watered from some time, and people came from near, and far to get their first glimpse of the large marsupial.

But after a few days the animal became very restless (know doubt wanting to get back to its own habitat), and one night jumped over the partition between the stalls, and choked itself, the chain not being long enough to let its feet touch the floor on the other side.

The late Mr Cotton was my neighbour after his retirement, and often told me his story—back more than forty years ago.

From Mrs Annie Jackson, West Wynyard

. . . As a girl of ten years old I took up residence with my parents at Rocky Cape. In that time thick Bush surrounded us. My father had cleared the Bush back about two hundred yards, and up the Back, We had the fowl house. Every now and then we were losing chooks, And the blame for this fell on the Tiger Cats or Devils.

Then one summer morning at about five thirty My Father after going outside, came back in, and in a whisper told my Mother and I to come quickly and quitly And look at this. Obeying his Instruction we moved to the Door And peeped around the Door. Prowling on the edge off the Bush up the Chook House was an Animal like a Big Mushroom coloured Dog. At this time he did not attempt to touch the Chooks. All of a sudden he came down toward the House.

We moved back into the Porch And he just passed by. But all of a sudden Dad Reckoned he got our Scent, And became aware of our presence. Because he took off, But we got a marvellous look at this light coloured Creature, with the Black stripes on his Back.

From Bruce Walker, Wynyard

I am probably one of few who can personally recollect having seen the last Tasmanian Tiger caught. The precise date I do not recollect, it having been recorded I believe at Mawbanna in the early 1930's. As I'm now approaching sixty this was before I had reached my teens.

This was during the depression years when many trappers eked a meagre livelihood snaring and trapping wallaby and possum or rabbits occasionally getting a tiger which they usually killed. There was however a local Wynyard identity Jimmy Harrison a collector of native fauna who found a ready market for the tiger at Australian Zoos, and paid trappers for live specimens.

Having secured the last one known, and which had a broken leg he called on my father Dr Gordon Walker to save it. He died in 1977 aged 86, having been a well known coastal surgeon for 1/2 century when Wynyard was the base hospital serving Penguin, Burnie, Wynyard & Circular Head municipalities.

I vividly recollect this seemingly vicious, snarling, tiger in its netted box which frightened me when I accompanied my father, however he soon quietened it. After covering the cage with bags he then threw in wads of cotton wool soaked in chloroform and it was quickly anethesised sufficiently for my father to splint its leg using wood and plaster. This process was repeated some weeks later to remove the splint and I believe the tiger was sent to Hobart Zoo.

From Gus E. Tubb, Malbina via New Norfolk

This is not intended to be a story of maybe but true facts. I saw my first tiger in the Hobart Zoo a long time ago. Since then I have seen it at Piper's River, Westerway, Fathers Marsh (Dry Creek, N. Norfolk) & Molesworth. The one at Molesworth has been here for over twenty years, which make me think there must have been more than one as I don't think they live that long. Last Feb. my son ran right into it crossing the road for water. You may remember it was very dry at that time & it was going to the creek for water. He—my son—was in his car & only a quarter of a mile from my home He rushed up home to ring the Wild Life so they could come and see for themselves. The answer was this WE DO NOT HAVE ENOUGH MONEY

TO BUY THE PETROL TO TAKE US UP There was nothing realy done about it. It was a golden chance as it had Cyclone fence three parts of the way around the block. A Police Stock Inspector advised me not to report that we see this one at home as it would probably be hounded away. Here is a few things you dont know. Nothing else in the bush has a bright green eye round in shape. At night with the light it looks nearly twice as big as normal. One other thing when it kills & is not real hungry it will open the animal & make a very neat job of taking out the kidney's & fat & leave the rest. I have seen this trade mark several times on things as small as a native cat. One other thing few people know is that the tiger cant resist hot bacon fat. Every time we cooked bacon at the snaring camp he would come right in close & scare hell out of the dogs. The cooking fat smells a long way. It is common talk in the district that the tiger was shot & wounded a couple of months ago, but since the Wild Life people treated us like it did I did not trouble to report same.

From Kath Williams, Devonport

This is a true happening I am writing about. It concerns a woman the name being Mrs Myrtle Whittaker who lived at Scottsdale on the 10 mile track Road leading to Springfield.

We move to Scottsdale in the year 1952. Mrs Whittaker had been lost for a week now when they found her she claimed she slept under a log with Tasmanian tiger cubs. Well for that.

But, around 1956 we moved out on the Ringarooma Road on the corner of the Ten (10) mile track Moris House who had a mill there so we got to know Mrs Whittaker well as she only lived about 1/2 mile up from us. She claim that the tiger was around her hut so she informed the press & they came up (I think from Hobart) they camped the weekend at her place with flash camera etc. hoping to see the Tasmanian Tiger. Now she came to us, and ask if she could have a lend of our Bitch Dog. She intended to lure the Tiger to the Dog (being a Bitch). We did lend her the Dog & she tied it up for the week end. But sad to say no Tasmanian Tiger turn up. She was so convinced in her own mind that the Tiger was still around & claimed she often saw it. But it must have been a dissappointment to the Camera men & Press when none came around. This is a true happening & if you probably looked up the records you will find it so. As for the exact Date I could not tell but they are the years or thereabouts. Both were in the Newspapers. Mrs Whittaker being lost & her having seen a Tasmanian Tiger around her place.

From Mrs Ruby Lorkin (89 years old), Chigwell
Over one hundred years ago, during a depression, my father the late James Price was forced to leave his wife & family at Hastings, S. Tas & went with two companions, prospecting down Savage River, in the rugged west coast.

There were no motor cars at that time, so they walked all the way, each carrying a swag including a camp-oven & tent.

They lived on native animals, such as rabbits & porcupines (which are very tasty when skinned and baked, like suckling pig).

One evening near dark, a striped Tasmanian Tiger came creeping around near their tent & ate the scraps of meat, bones etc from the scrap-heap where they threw out their waste food etc.

The men watched from inside the tent, as she came every evening.

After about a week they noticed she was taking the food away in her mouth. They ventured outside, but could not get close to her, or tame her.

They followed her, keeping a safe distance behind, she led them to a large fallen tree, & underneath this, saw she had three young pups.

She glared, & growled at them as she saw them coming near her Puppies, so they decided not to approach any further, but she did not attempt to rush them, or bite them.

I think dogs & other animals can sense who are friends, & who enemies. the dog continued to come, each night to the camp-site, for food for her young ones.

The men left for home soon after this.

My father said he did not think Tigers were a savage animal any more than a house dog, which has gone wild, & will only attack sheep, & lambs etc when hungry.

I well remember seeing the last Tas Tiger at the Domain Zoo, in Hobart. They also had a tea-room there at that time, & it was a very popular walk for tourists, & Hobart citizens to go out there on sunday afternoons I think, if there any Tigers still living, the rugged Gordon River of the North West Coast would be the most likely place to find them. Of course, at the time I write about there were hundreds in the bush, & they were shot down, indiscrimately, by hunters, etc,

Its a pity we have lost so many of our Native Animals in this way. It is the same with Rabbits, Now they are talking of farming them for meat consumption.

We wake up too late.

The year 1985 is an important one in the thylacine story. Eric Guiler's *Thylacine: The Tragedy of the Tasmanian Tiger* was published, bringing together for the first time virtually all of the known scientific data on the animal. Guiler—once a zoology student of Professor Thomas Flynn—had exhaustively researched the Woolnorth records and interviewed many old-timers with thylacine experience. There is no indication that the book's timing was deliberate; the following year, 1986, marked the fiftieth anniversary of the death of the last known thylacine. In zoological practice a species is officially deemed extinct if it has gone unrecorded for a period of fifty years. Guiler's theories and conclusions represent a major landmark in the thylacine story. His preface ends: 'I am still searching.'

Two years prior to the Guiler publication, Tasmania's extraordinary Franklin River dispute was decided by a High Court ruling which enabled the federal Labor Government to kill off a state government-backed hydro-electric scheme to dam the river. Many development-versus-conservation clashes were to follow, with the thylacine to the fore when the long-simmering Tarkine dispute erupted. The Tarkine, in Tasmania's far north-west, is Australia's largest continuous area of temperate rainforest, spreading across the almost totally unpopulated areas of the Donaldson and Savage River catchments. Originally known as the Arthur Pieman Protected Area (the Pieman River is reputedly named after cannibal convict and escapee Alexander Pierce, who was a pie-maker), the Tarkine was renamed after its Aboriginal inhabitants, the Tarkiner people. The area has many sites of Aboriginal significance. Early in 1995 the Liberal government revived plans for bulldozing a road through it to link Smithton on the north-west coast with the mining town of Zeehan, north of Macquarie Harbour. Promoted by the government as a future tourist enhancement, it was dubbed by conservationists 'the road to nowhere'. Its opponents said it was being built solely in the interests of forestry and mining companies.

The unsuccessful campaign to save the Tarkine operated at various levels. A direct action group, the Tarkine Tigers, came into existence (their first meeting was held in a room above

Knopwood's Retreat), which took the thylacine as its symbol. Locking-on to bulldozers and other equipment became a common tactic, requiring police to cut the activists free. The giant 'Save Tarkine Wilderness' banner with its drawing of a thylacine would then be draped in front of the inert machinery, while hours of road-building were lost. To many Tasmanians the Tarkine Tigers truly were feral, unpredictable, dangerous—a strange echo of the groundless fear the real thing had engendered in the nineteenth century.

There were many arrests. One arrestee, Bob Brown, spent five days in custody. He drew attention to yet another dimension of the conflict in commenting upon the new Regional Forest Agreements drawn up between Canberra and each of the states:

> It means that if the Tasmanian tiger emerges out of the Tarkine, along the pipeline track, for instance, and it is then decided you need 10 000 hectares to protect it and any other animals in there with it, then the Commonwealth [government], if it does that, is obliged to compensate [the forestry company] for the loss of that resource—all those trees. It would cost tens, if not hundreds of millions of dollars.[4]

Conservation issues remain highly emotive in Tasmania because there is seen to be a close correlation between one form of environmental extinction and another. For many Tasmanians, large-scale old-growth logging was so comprehensive for so long that it differed little from the thylacine slaughter in which their forebears indulged. Thus it was that in the lead-up to the 2002 state election a pressure group, the Tasmanian Community Alliance, was formed to put the extinction issue firmly on the agenda. The Alliance's television advertising campaign was spearheaded by the famous/infamous 'Mr Weaver bags a tiger, 1869' photograph, with the advertisement drawing 'a parallel

A pressure group, the Tasmanian Community Alliance, used this pamphlet during a state election to gather support for its stand on protecting old growth forests. The insert photo—'Mr Weaver Bags a Tiger'—is famous for being infamous as a symbol of the wanton persecution of the animal. (Tasmanian Community Alliance, Hobart)

between old-growth logging now and Tasmanian government policy 100 years ago which allowed Tasmanian tigers to be shot until they were extinct'.[5] The advertisement went to air regularly in prime time. Geoff Law, longtime campaign coordinator of the Wilderness Society, was unequivocal about the reason for its success and what might flow from that:

> The Thylacine is a potent symbol of all the mistakes of the past, especially in the imagery of a dead thylacine upside down in the Weaver photo. There's a sense of tragedy, vulnerability, the immense regret that people experience. If only we knew then what we know now, that wouldn't have happened, and yet we admit cheerfully to turning these magnificent, statuesque forests into woodchips for export . . . If there's an end to old-growth logging then we'll have a proper World Heritage area with the most appropriate boundaries for it, instead of a

> hacked-up ragged edge that we've got now, with chunks left out for loggers and miners. In terms of an all-pervasive, oppressive issue, the forestry one will hopefully be the last one to be dealt with . . . They don't need to keep invading valleys that have never seen an axe.[6]

The thylacine had only once before been used to so directly attempt to influence Tasmania's political landscape. Its last appearance as a political animal had been in the 1887 debates which put the statewide bounty in place. More than a century later, it continued to vex Tasmanians. In the words of Geoff Law:

> It's funny how there's this fascination with the creature, partly because it's a large predator, partly because of the mystery as to whether it's there or not. And it even becomes a joke, almost. 'The Thylacine: maybe we'll see one' . . . It becomes light-hearted, which completely glosses over the tragedy that occurred. The thought that it might still be there actually allows people to overlook the horrific crimes of the past in ensuring its extermination. A quite deliberate, systematic extermination of the thylacine occurred, but people don't want to face up to it—that that's what Tasmanians did.[7]

14

THE TIGER IN COMMERCE AND ART

> I was driving home to Railton . . . in a trap pulled by a fine white pony called Robin. As we neared Railton he suddenly stopped dead in the road so that we were nearly thrown out. He refused to budge and then we saw the reason, a Tasmanian Tiger had come out of the bush to cross the road and both tiger and pony met in the middle and eyed one another . . . The tiger slowly crossed to the other side and disappeared into the thick bush. It was dusk but we had a good side view so that its stripes were distinctly visible. It was several minutes before the pony would move on.
>
> ELMA KIDD, WEST LAUNCESTON

The longer the thylacine stays dead, the greater the interest it arouses. That is quite an achievement for an officially extinct species which in its lifetime was not seen to have the fearful charm of the exotic jungle cats, Africa's big five, grizzly bears and the like. Yet it has been kept alive by considerably more than regret. As symbol, art object and metaphor, the thylacine holds lofty status, being accorded a degree of patriotic regard (like the bald eagle), a degree of reverence (like the cat in ancient societies), achieving something of the symbolism of the unicorn in literature and heraldry, and becoming a sporting icon (in company with wallabies, kangaroos, devils, springboks, lions, cougars, hawks, magpies, dolphins, sharks and jack jumpers, the island's highly toxic endemic ant, after which Tasmania's basketball team is named).

The thylacine has about it a sufficiently powerful and ambiguous mystique that it is able to comfortably represent and embrace much that is Tasmanian. Thus, in a display of unintentional dark humour, a pair of heraldic thylacines stand rampant in the state's Coat of Arms, supporting and protecting a ram in the centre of the shield. It may also be seen on Tasmanian car licence plates and is the fetching logo of the City of Launceston. A television station, a furniture removal company, a fruit and vegetable wholesaler, an abalone company, a bus line, a supermarket, a publisher, craft stalls at Hobart's famous Salamanca Market, retailers of tourist kitsch, and many others, merrily trade under and off the tiger identity.

Tourism Tasmania, operating within a government department, makes full use of the abstracted images. The original

This former Tourism Tasmania logo abstracted the thylacine to a high degree that was appropriate given its extinction status. Even so, as a brand symbol it was an undoubted success, attracting large numbers of tourists to Tasmania. The similar current logo emphasises the abstracted face, over the words 'Explore the Possibilities'.
(Tourism Tasmania)

logo's designer, Ian Kidd, explains the philosophy behind it, reading in part:

> The Department, having commissioned intensive research and design activity, [now] had a visual identity with which to identify and market Tasmania to target audiences locally, interstate and overseas . . . Whether still alive or extinct, the Tasmanian tiger is unique and immediately identified with the State to the extent that it has become mythicised. As an animal it isn't just a symbol of Tasmania; as has been depicted in the logo, it is a symbol of the Tasmanian Experience . . . The symbol graphically depicts a Tasmanian tiger in a wilderness environment. The form comprises three elements; being the tiger, vegetation and water composed within an imaginary circle.[1]

In the non-government sphere, Hobart's Cascade Brewery Company reigns supreme in commercial usage of the thylacine. The gracious sandstone building and storage tanks of Australia's oldest brewery, established in 1834, occupy an unrivalled site at the base of Mount Wellington. That striking image long carried much of the company's advertising fortunes, but a point was reached in the 1980s when the company needed a different kind of help. Enter the thylacine.

A growing demand for premium brews, nationally and internationally, obliged Cascade to test that market, in the face of stiff competition from the big established players like Heineken, Budweiser, CUB's Crown Lager and Carlsberg. Cascade's new product, released in 1987, was a considerable success, not least through the use of the mountain's pure water, locally grown and malted barley and a long maturation period of six weeks. But in the crucial marketing area lay its real secret ingredient: tapping into Tasmania's new image as environmentally clean and progressive, a complete contrast to the jaded metropolitan world of the target market, well-heeled beer drinkers.

Designer Ken Cato created a label closely based on H. C. Richter's 1850s lithograph for John Gould. (Gould had written: 'The circumstance of a fine pair, male and female, of the *Thylacinus cynocephalus* being now living in the Gardens of the Zoological Society in the Regent's Park, enables me to give the best figure of the animal that has yet appeared . . .'.[2])

In combination with an unusual light green glass and expensive packaging (the latter to protect the beer from hop-spoiling ultra-violet light, which doesn't penetrate brown glass), the visual strength and somewhat mysterious allure of the label helped make the new premium line instantly successful. The following year Cato won an award from the New York Art Directors Club. Cascade's Premium Lager and Premium Light became bestsellers throughout the developed world.

In all likelihood Cato's design was inspired by the fact that the brewery had had a prior relationship with the thylacine. In the 1870s artist William Piguenit used one in an advertisement he designed for the company; also at that time, stonemason Edward Martin Richards sculpted a thylacine atop a beer barrel for the company's city office. It's a well-known Hobart feature—for those who know to look up at 156 Collins Street.

Even here there are controversies, though one is mild and amusing. For some reason Cato, in adapting the Richter lithograph, extended the stripes to the tips of the tails. It is a sufficiently odd alteration to make one want to go and buy a sample of the product, if only to gaze at those most unthylacine-like appendages.

At a different level of controversy altogether an Aboriginal legal bid sought to claim copyright over native animal images—particularly the thylacine, emu, platypus and kangaroo—as sacred symbols. According to Tasmania's Aboriginal and Torres Strait Islander Commissioner, Rodney Dillon, in an article in *The Mercury* on 31 January 2002, 'These native animals are part of our people and it is an insult the way they are being used . . . They steal our land and they steal our animals and then they use them as their signs. Cascade should use a fox on its labels instead. That's what they [whites] brought with them'.[3] A follow-up report the next day stated that 'one senior Cascade source' conceded that Aboriginal people had a point and that the company recognised the sensitivity of the issue.

Cricket is Australia's national summer game and the thylacine features prominently as the identifier of Tasmanian cricket. The logo of the former Tasmanian Cricket Association (TCA) cleverly integrated the animal with the game by having its stripes represent the seam of a ball. The state men's and women's cricket teams are both known as the Tasmanian Tigers, with a heavily stylised tiger face projecting the right degree of sporting alertness and aggression.

The former TCA's Media and Public Relations Manager, David Boon (himself a Tasmanian icon, with a test cricket career aggregate of 7422 runs in 107 Test matches at a splendid average of 43.65), explained the 1995 adoption of the name Tasmanian Tigers:

> Its purpose was to identify and brand the Cascade Tasmanian Tigers cricket team as distinct from the corporate logo of the Tasmanian Cricket Association. Informal but extensive research showed overwhelming public support for the Tiger image. The research had involved seeking the opinions on a range of names from high school students—a key market sector for merchandising—players and the general public. Ten names were offered.

> The Tigers was a clear winner with thirty-six per cent of the vote . . . The Tasmanian tiger was used because of its uniqueness. No other state can claim it. It is/was a sleek, cunning and aggressive carnivore—a killer. If it still exists, it is elusive, surrounded by mystery and extremely hard to track down. It certainly projects an appropriate image for our cricket team. The logo is associated with all that is the Tasmanian Cricket Team and it is used on all advertising, uniforms and merchandising relating to that team.[4]

The above three examples, prominent in their respective areas of tourism, commerce and sport, clearly demonstrate the financial worth of the animal. What, one wonders, would struggling early Vandemonians—free, or wretched in chains—have made of that? It is beyond the scope of this book to calculate the earning power of the thylacine but, as the face of Tourism Tasmania alone, it's up there with the primary industries.

To many, however, the thylacine has nothing to do with tourist dollars, brewing or sporting excellence. The 1991 painting *Got the Bastard* (see colour plate) by Tasmanian Aboriginal artist Karen Casey, held in the National Gallery of Australia, is a large and imposing work, measuring 2 by 1.5 metres. The work directly references the famous/infamous Weaver kill image (see page 202). In personal communication with the authors, the late artist's son Daniel Young wrote: 'All my conversations with Mum about the work were always about her take on the Weaver photo that sat in prime position at TMAG—then the comparison of the last photos of Truganinni, the sadness and the darkness in these images—the end was near. Karen wanted to capture the aggressor in all his satisfied swagger and with the tiger almost disappearing into the background. It's a very imposing work and I would say her strongest painting.'[5]

Assessing the painting in 2012 Latrobe University anthropology lecturer Dr Nicholas Smith wrote:

> Extinction is the ligature that binds Palawa and thylacine in the Australian imaginary. The settlers' maladaptation to the environment and separation from the indigenous 'other' achieves symbolic inversion in the Tasmanian tiger. This inversion is powerfully captured in . . . *Got the Bastard*, an explicit depiction of the Tasmanian settler's conquest of colonial nature and native . . . Casey's disturbing composition shrewdly juxtaposes the devastation of Tasmanian biota with the destruction of Aboriginal Tasmanians . . . It is significant that the thylacine in Casey's painting hangs upside-down, partly mimicking colonial images of the thylacine and because structurally, the mythological Truganinni and the mythological thylacine invert one another. Thylacine narratives posit existence in the face of extinction, whereas the myth of Truganinni posits extinction in the face of existence.[6]

A 2021 *Sydney Morning Herald* retrospective of the late artist by Adrian Newstead reaffirmed the work's angry vitality: 'From the outset, works like *Got the Bastard* . . . challenged notions of what Aboriginal women's art was about. The brutish overbearing redneck with a rifle across his lap and the extinct Tasmanian Tiger haunting the background was an incarnation of the ignorant men who annihilated the Tasmanian Aboriginal population. While Casey's strong expressionistic treatment concentrated the viewer's attention on his raw powerful arms and the slash of his wide mouth, his huge eyes would not meet your own.'[7]

For some artists the thylacine is the centrepiece of their work and career. Melbourne artist Daniel Moynihan has been incorporating the thylacine into his paintings for over three decades, in the process depicting it thousands of times. He tends to use settings alien to the animal, which renders his imagery more dramatic. An example is his large 2001 canvas *The Meeting—Wickerman and the Tasmanian Tiger* (see plate section), in which

> he has brought together his ancestral Celtic past and joined it with the myth of the thylacine. According to the Celtic myth, before crops were planted the Celts and later pagans would build an effigy and then burn the Wickerman, gather up the potash it left behind and spread the fertiliser on their fields for a successful season of crop production. The destruction of the thylacine was also considered beneficial for the growth of the early colonial farmers and its sacrifice a small price for the overall benefit of the community . . . The thylacine, the Wickerman, the cityscape, the convict tools of labour have all been collected and placed within the canvas embodying thirty years of memory and the representation of myth and its associative value to the individual and the society which it inhabits.[8]

By contrast, Moynihan's contemporary Michael McWilliams, who lives near Launceston, generally places his thylacine in a rural setting. But there the reference to its natural state ends and, like Moynihan, McWilliams gives life to the animal as a form of penance for what has been done to it.

McWilliams ran an antique business in Longford. Many of his works are painted on furniture, or wood panels, reflecting both his appreciation of the material and the fact that it is accessible. In particular he paints on colonial furniture, arousing in the viewer an odd sensation through the placement of his watchful acrylic thylacines on the meat safes, cupboards and boxes that were crafted when the animal was very much alive.

His paintings, according to the catalogue notes for a 2001 exhibition,

> have a perverse juxtapositional quirk of some kind where the presentation is not quite as expected. A persistent half-real, half-imagined, surreal sensibility runs throughout . . . Rural scenes are framed in painted borders, animals are presented on pedestals as monuments or they stare out with the considered and conscious presence of formal portraits. The stare is always engaging and often challenging; we cannot help but become involved in the directness of the exchange. Is it a look of confidence, of accusation or detachment? Even when not the main subject of the painting, the elusive but determined thylacine has the presence of a figure in a tableau. It lurks like a prompt for both consciousness and conscience.[9]

The major work from that exhibition, *Up On A Pedestal* (see plate section), is particularly symbolic, showing the animal in its desired terrain (felled logs excepting):

> Where once a bounty was collected and tanned skins sent to the United States for waistcoats, a thylacine now stands on a stone monument, mounted in a magnificent tramp-art frame (that has waited over ten years for the right subject). It reflects—too late—the extraordinary shift in status, from pest to pedestal, that we have conferred upon this animal in the recent decades of our living memory.[10]

Such is the wealth and variety of thylacine visual imagery that it was the subject of a 2005 PhD thesis researched by Carol Freeman at the University of Tasmania's Literature and Environment Unit. (Thesis title: 'Figuring extinction: Visualising the thylacine in zoological and natural history works, 1808–1936'.) The unit was headed by well-known academic and writer Peter Hay, whose first published story, *Lagunta*, was about thylacines as both hunters and the hunted:

> Nta slid through the mottled moonshadow of rainforest, the attendant kindred matching his silent tread. Unseen they were, but palpably there at each flank; eldritch stone-eyed hunters threading a noiseless path through the tangle of moss and log . . . In the fire's halo stood the two strangest creatures Nta had ever seen. They were enormous; huge erect monsters, standing easily on their hindquarters . . .[11]

Freeman's thesis is motivated partly by a desire to contribute to information about the role of representations in constructing, perpetuating and changing the perceptions and actions of humans toward animals:

> To make sense of the multitude of thylacine images and the accompanying narratives and discourses, I focus on groups of images in a particular medium or representational space. These include zoological illustrations . . . photographs and films of killed and captive animals . . . logos, emblems and coats of arms . . . advertisements . . . artworks . . . Some images seem to encourage destruction of the animal, others assume or ignore the probability of extermination, while many interrogate the possibility of death or resist the notion of loss. At present, artworks and advertising are emerging as sites of vigorous resistance to extinction . . .[12]

In yet another medium, a 2001 CD-ROM *The Tragedy and Myth of the Tasmanian Tiger* was artistically and technically creative and comprehensive. The company responsible, Hobart-based Roar Film, coincidentally happens to have a cartoonish roaring thylacine as its logo.

The interactive nature of that medium made it well suited to the telling of a story such as the thylacine's, where a chronological linear history is but one of many aspects relating to the animal.

Thus, there are about six main entry points of equal weighting—biology, distribution, extinction, and the like. Taking one of these as an example, 'Spirit of the Tiger' invites the user into a further range of topics—commercial value, the thylacine in art, searches, vampire myths and more. The prompts are visually enticing, being objects such as a painting on a wall or the drawers of an old filing cabinet in the cluttered rooms of a colonial-style house. The work is rich in imagery, music and commentary from contemporary thylacine experts.

Steve Thomas, creative director and co-writer of *The Tragedy and Myth of the Tasmanian Tiger*, moved to Tasmania in his late twenties and, like many others living in the country, used to drive the pitch-black winding rural roads at night wondering if he might just see a thylacine—until he reached an inevitable conclusion:

> I believe that thylacines are extinct. Every credible authority I've spoken to believes that. I know there are some extremely knowledgeable bushmen out there who believe in it, but I do think people are grasping at straws. It's very unlikely that once a species gets below a critical mass it can still sustain itself. It's very likely that the species didn't become extinct in the 1930s but in the 1960s or 70s . . . The debate has become much less about whether the animal exists and more about its demise and why we now find ourselves in this position where we're making art and writing books and talking about this animal which no longer walks the planet. And I think that's an interesting and welcome change. Talking for instance to the people at the Tasmanian Museum and Art Gallery, I discovered that their focus is now almost on a kind of natural history and forensic archaeology. Using what they've got, they're trying to piece together stories of the animal while it was still alive. They have a wonderful

> archive of material and I think that their work, and Paddle's, has contributed to a new consciousness about the animal. One of the best recent uses of the thylacine was in the anti-old growth forest ad where the question is posed: do we want this to happen again? That's a real progression from where we were twenty years ago.[13]

The first book devoted exclusively to the thylacine—*Search for the Tasmanian Tiger* by Quentin Beresford and Garry Bailey—had its own interesting story, as related by Bailey, a former editor of *The Mercury* newspaper (which of course has its own long tiger connection):

> The idea for the book came from my brother-in-law Quentin Beresford . . . and his then girlfriend Sue Dyson. The idea bumped along without coalescing into a firm idea of where it was going. Because there was little or no co-operation from the then thylacine expert Eric Guiler, who was working on his own book, they were limited in their scope . . . I suggested they limit it to the relationship between man and thylacine and target a market that might actually bring sales . . . I [also] made a few suggestions about structure, snappy intros and some racy prose. About the only thing I left out was a love interest! Eventually it ended up in bits at my home so I basically wrote it using their research. I did more of my own sleuthing using newspaper libraries here and interstate, interviewing people and organising pictures from our archive and the Tasmanian Museum and Art Gallery. Quentin meanwhile approached Dan Sprod of Blubber Head Press . . . [who] took the manuscript and actually added to it with some research of his own, including finding the actual date of the death of the last Tasmanian tiger in captivity. That alone corrected the popular record, which was a year out. Dan decided on a soft-cover book (or booklet, to be more correct) and went for a print run of what I

> believe was 5000 copies . . . It sold like mad. I . . . still receive a very modest stipend through its use in libraries and the photocopying. I did have a strong personal interest in the thylacine, having done quite a few stories as a reporter on sightings and having had regular contact with James Malley the renowned tiger hunter and Bob Brown who also conducted a search with the mercurial Jeremy Griffith. However, my main motivation was proving to my argumentative brother-in-law that you could write a popular history and make a quid. The most interesting recollection is that the release of the book seemed to spark an increase in tiger sightings and I received lots of calls from people wanting to relate their experiences . . . Does the tiger exist? No. But, as always, that's head ruling heart. If it does exist I want the picture first.[14]

The thylacine is sparsely represented in adult fiction. Unlike a visual creative image, depicting it imaginatively to a mature reading audience has built-in limitations and traps for the unwary author. Still, there have been a number of novels—accompanied by, inevitably, controversy. Erle Wilson's *Coorinna: A Novel of the Tasmanian Uplands* (Melbourne University Press, 1963) is distinguished by being one of the first thylacine-specific novels; it would be hard to find today and perhaps worthy of a tiger-style hunt for that reason.

Nan Chauncy was an eminent Tasmanian children's author with fourteen published books. Her illustrated *Tiger in the Bush* won the 1958 Australian Children's Book of the Year Award. ('A Tasmanian boy reveals the habitat of the almost extinct Tasmanian tiger to someone who may want to capture it, and now he must find a way to protect the animal.'[15])

Since the 1980s there have been a good number of works of fiction published in the junior and young adult categories, as well, either about the animal as it was or about its present mysteriousness and felt presence—both of which are exploitable fictional territory, though perhaps easier to shape for juvenile markets. Thus in Beth Roberts's *The Magic Waterfall* (Rainbow Books, 1990) for the pre-teen reader, marsupial facts, extinction facts and the problems of inbreeding in a remnant population are openly presented but the parallel imaginative tale holds sway, as ageing Winnie Wombat helps Nenner the young Thylacine return over the mountains to his partner Corinna. Cleverly resolving the irresolvable, Roberts has Winnie render Nenner invisible through the agency of her magic moonlit waterfall. (There are also touches of humour—Winnie remarks condescendingly: 'Most wombats would not go out of their way to help a carnivore'.)

At the other end of the teen market, Melbourne author Michael Hyde's *Tyger Tyger* (Vulgar Press, 2001) is about a seventeen-year-old

> young gun footballer, following in the sprig marks of his late father and doing so by playing with the men of the Ballantyne Tigers instead of the boys. The coldly omniscient talent scouts have spotted him, but for dark mysterious reasons he can't always play at his best . . . The best parts of the book are set in Tasmania where Johnny goes to get over a hammy and also manages to unravel the mysteries of his recurring dream featuring a Tasmanian tiger and to better come to terms with his father's death.[16]

Like *Tyger Tyger,* Heather Rose's adult novel *White Heart* (Anchor, 1999) also works the thylacine into themes of loss and understanding, though as an adult novel it operates at a different level of complexity. Here the narrator Farley, who grew up in a Tasmania she loves, seeks inner understanding through a series of visits to the United States, where she participates in American Indian spiritual rituals: sundancing, sweatlodge purifying, praying and fasting. The thylacine connection, likewise, is mystical but very strong. While Farley moves between the United States and her Melbourne bookshop, her wise, feral-like brother Ambrose camps out in the wilds of Tasmania looking for thylacines. But neither are in fact there. ('Both extinct. Both mystical.') It's a tragic and moving novel, considerably underrated.

Rose says of her novel that it

> emerged at the end of a very intense part of my life. I think if I believed I would only write one novel in my life I would have set out to write a book that journeyed beyond the everyday into the realms of magic and spirit. A companion to our private quests for a deeper knowing of ourselves. Strangely, *White Heart* achieved that but only in retrospect. At the time it was very much a story unravelling. I was as surprised by it as anyone . . . To live in Tasmania is to live with beauty. But it is also to live with the threat of destruction and loss. The tiger illustrates that but so too has the past decade of forest practices. If Tasmanians are demonstrating a new Western sense of environmental consciousness it is because rivers, sky, mountains, plains, sea, wildlife and forests have quietly taught us the concept of stewardship. To harness that teaching is to tap into a powerful source of magic, as all indigenous cultures know.[17]

Julia Leigh's *The Hunter* (Penguin, 1999) became an international bestseller while attracting considerable flak. The storyline is as minimalist as its central character, M, a man with SAS-style hunting and survival techniques who is sent to Tasmania by a biotechnology corporation to find and kill a thylacine and take samples of its DNA for military purposes.

He is successful in his mission. The fate of the thylacine? He buries it. (In a perverse real-life twist, rumour had it that some Tasmanian hunters let it be known that if they discovered and killed a thylacine they'd bury it and tell no-one, to prevent their shooting territory being locked away. Perhaps it's already happened, lots of times!)

The novel was widely praised by UK and US critics, for its spare prose and philosophical overlay in the manner of a Moby Dick quest, pitting man against nature. It also drew comparisons with Ernest Hemingway's *The Old Man and the Sea.* But closer to home, and particularly in Tasmania, Sydney-based Leigh's use of the island was thought by some to be both gratuitous and inaccurate. It led to a debate about the appropriation in fiction of a place with which an author is but passingly familiar. For former University of Tasmania reviewer Andrew Peek, '*The Hunter's* setting, in rural Tasmania, pits hippie and conservationist against backwoodsy yokels. The bush is by turns impassive and threatening. In the best tradition of Tasmanian Gothic, it is a place of ghostly presences, beautiful, eerie landscapes and "devils" . . .'[18]

In 2011 *The Hunter* was made into a successful drama film of the same name, starring Willem Dafoe, Sam Neill and Frances O'Connor, with an animatronic thylacine making an appearance at the movie's climax.

Tasmanian discomfort with and dismissal of Leigh's novel became, in some instances, outright attack when Chloe Hooper's *A Child's Book of True Crime* (Random House, 2002) hit the bookstands. Like Leigh, Hooper is not Tasmanian—she lives in Melbourne—and her novel was also snapped up internationally (while still at manuscript stage by New York super-agent Andrew Wylie) and published in great numbers to general acclaim. In this story Kate, a young teacher in a fictional town near Port Arthur, has an affair with the father of her precociously brightest pupil. His wife is the author of an acclaimed true-crime book on a foul local murder committed twenty years previously, which ominously involved a young mistress. Then strange, rather spooky things start to happen to Kate, and . . .

The thylacine element in this novel is satirical and tangential to the actual storyline, in the form of a parallel children's story in which a crime is investigated in the animal world by Terence Tiger, Kingsley Kookaburra and others (move over, Beth Roberts) but the tiger usage firmly roots the novel in its 'Gothic Tasmanian' setting, as does its location in Port Arthur, scene of the 1996 slaying of thirty-five people by a deranged Hobart gunman. The front cover of the first hardback edition of the novel has an innocent, cartoon-style drawing of a group of Tasmanian animals, with Terence Tiger the thylacine prominent.

15

CLONING

> My mother well remembers many years ago when her brothers brought home a young Tiger (they had caught the mother in a snare) & chained it in a corner of a room in the house for about a fortnight. They fed it on kangaroo and wallaby meat, & she says it used to growl & snarl if anyone went near while it was eating . . .
>
> MRS OLIVE SIMMONS, SCOTTSDALE

Why clone a thylacine? The answer, in keeping with the contrary phenomenon of the animal, is both complex and divisive.

Artificial manipulation of flora and fauna to suit human requirements is an ancient practice, dating back to the domestication of wheat and subsequent selective cross-breeding for improved yields. The domestic dog has been interbred to achieve specific characteristics: tenacity, docility, endurance, speed. Chimeras—such as the liger, the offspring of a lion and tiger—while the product of natural reproduction, are unusual in that they are the product of two genetically different species. In all of this genes are the key.

A gene is a unit of deoxyribonucleic acid (DNA, sugary complex molecules) and strings of genes are arranged, or sequenced, on the chromosomes of a cell's nucleus. Genes pass on specific characteristics from parents to offspring. The full set of chromosomes of an individual is known as the genome. (The human genome has between 30 and 40 000 genes.) It follows that the ability to isolate and remove the genome from a plant or animal is the first step to artificially recreating that organism identically (cloning).

The second step in cloning is to ensure that the replicated genome is perfect. The third step is to grow it—in the case of animals, through the otherwise natural procedure of gestation in a (surrogate) mother.

In 1999 the Australian Museum began a ten-year project to clone a thylacine, led by the Museum's Director Mike Archer, a palaeontologist and conservation biologist with a particular interest in extinct species. His PhD was on Australia's carnivorous marsupials and he has since done much in the palaeontological line, including discoveries of seven ancestral thylacine fossils in Queensland, 'in about four genera [closely related species], only one of which is *Thylacinus*'.[1] His credentials as a thylacine expert and devotee are unquestioned. He has dated multiple ancestral thylacine species to about 25 million years ago, at which time they differed considerably in size, until by about five million years ago when just a single species remained.[2] When he came across a preserved thylacine pup in the Museum's collection many years ago, he paid it more than passing attention, later reasoning that because it had been kept in ethyl alcohol rather than formalin (since 1866), its DNA might not have been destroyed and therefore—at some future point—might successfully be extracted.

Seed funding for the ambitious project was provided by the New South Wales government and a private trust. Obtaining money was in all likelihood the easiest part of the equation.

Certainly, the pitch for funding must have sounded good: 'What Mike Archer and his team are attempting is as scientifically exciting and technically challenging as splitting the atom or landing a man on the moon'.[3]

Significantly trickier was the question of the DNA itself: acquiring it from a variety of sources, successfully extracting it, hoping that it would be undamaged, replicating it. Thylacine remnants are scattered in museums and other collections around the world, legacies of the sixty-plus animals known to have been displayed in zoos between 1856 and 1936. Obtaining access to such rare and valuable material, for the sake of genetic variety, would prove crucial. Did the Australian Museum have access to it? 'Yes. Most of this knowledge about other specimens has been co-ordinated via the Tasmanian Museum and Art Gallery in Hobart.'[4]

Thylacine DNA is aged, fragmented and dehydrated. Its reconstruction is ferociously complicated, even when explained simply:

> The ethyl alcohol in which infant thylacines have been stored preserves tissues and genes, but in doing so, water, the very substance which is vital to life, is expelled. Scientists would need to rehydrate not only the DNA in the thylacine nuclei, but the entire matrix of nuclear proteins and enzymes that maintains its structure, and oversees its orderly replication as the cell divides and multiplies to form an organism. Therefore, more is needed than just the coded instructions for how to build a thylacine. Obtaining the complete package of necessary biochemical elements will be the determining factor in whether thylacine cloning is a success or failure.[5]

The Australian Museum team set to work and DNA—not without difficulty—was isolated and extracted from the 1866 pup and subsequently from two other pups. The tissue sources included bone, bone marrow and muscle. The extracted DNA, however, represented only a miniscule percentage of the complete genome of the specimens.

Three years into the project a significant breakthrough was announced. Through a process known as polymerase chain reaction (PCR), four thylacine genes were successfully replicated and copied into millions of short fragments of undamaged DNA.

The next stage was to involve the construction of synthetic chromosomes from the fragments, building towards the perfectly replicated genome. Archer referred to it as creating a 'genetic library' of a thylacine's DNA. (The metaphor is apt. The Library of Congress has holdings of many millions of items, each one with a specific classification assigning it to one and only one place on a shelf.) How would the library be built?

> The concept is that the DNA of the thylacine is tagged onto the DNA of living creatures (plasmids of yeast or bacteria) which then look after the DNA—and this is the library. We are recovering DNA from three individuals. Whether that means each will have a library of its own, I couldn't tell you. This is an issue that the geneticists are better able to answer.[6]

Professor Marilyn Renfree went some way towards providing an answer. In February 2002 she was one of four experts participating in the ABC TV science programme *Catalyst*, discussing the thylacine cloning issue. Renfree, a University of Melbourne Department of Zoology specialist in marsupial reproduction, cited difficulties in cloning with living, undamaged material. Thus, Dolly the Scottish sheep, cloned in 1996, while an undoubted scientific achievement, had malformed organs and died suddenly in 2003. For Renfree, cloning is 'extraordinarily difficult . . . Dolly was one egg in 277 fertilisations, so there were 277 nuclei put into 277 enucleated eggs and we got one Dolly'.[7]

On that same television programme, Dr Ian Gunn of the Animal Gene Storage Institute of Australia was even more sceptical: 'To construct a nucleus and the DNA is virtually asking someone to construct a new species. To me, impossible'.[8]

A counter-argument was that science of this type could but improve. Another was that 'big science'—meaning a big story—attracts money that otherwise wouldn't go the way of science. And Archer himself was forthright: 'Personally, I think this is the most exciting biological project that's going to occur in this millennium'.[9]

Use of a host to build this library apparently posed no problems: 'If a thylacine is reconstructed and brought to life through this process, using a surrogate such as a Tasmanian devil, it will be a thylacine—and nothing but. With only thylacine nuclear DNA to build the animal from, it can't be anything else.'[10]

Even if that were so, the required use of a host presented problems. At that point, geneticists had not been able to grow nuclei taken from one living animal in another that was related, because the host oocyte (female germ cell) consistently rejected the implanted nuclei.

It would be a miraculous birth, a clone from aged DNA. It would, no doubt, 'be' a thylacine physically. But would it have an inherited 'personality' from the lineage of the 1866 pup? Or, like a captive-bred animal, might it lack a range of acquired behavioural traits as learned when wild-raised?

> This is a bit of suck it and see. I am convinced that a lot (most) marsupial behaviour is 'hard-wired' and under the control of the genome. This is the reason that behaviours of individuals raised in captivity commonly closely mimic the behaviour of individuals studied in the wild. But, at this point, this is speculation.[11]

It was proposed that a minimum number of thylacines would be cloned, and that natural reproduction would then take place followed by release into a safe area. 'No-one is clear yet about where the best place would be. We'll be leaving that up to the Tasmanian Parks and Wildlife Service, if the project is successful.'[12]

It's worth remembering that in 1967 the Fauna Board under Eric Guiler selected Maria Island as a sanctuary, should thylacines ever be found. The uninhabited island still holds that status, a fact of which Col Bailey reminded his audience when he gave a talk at the Tasmanian Museum and Art Gallery in September 2001.

Research into cloning extinct and endangered fauna had been going on elsewhere, and meeting with criticism as well as support. In Australia, the Mike Archer-led thylacine project met with five main areas of criticism.

Many within and beyond the scientific community doubted that it could be done at all:

> While managing to get PCR to work on a degraded source of DNA like a preserved pup is commendable, this breakthrough represents the tip of the iceberg in terms of creating a living, breathing thylacine . . . The 'hurdles' of decoding the correct sequence of DNA, constructing synthetic chromosomes and transferring chromosomes into a cell are tasks of such mind-numbing complexity as to make the PCR amplification of a few genes look like child's play.[13]

The cost of the project inevitably led to complaints that research efforts could be better spent elsewhere, not least in slowing Australia's alarming rate of extinctions. There was also scepticism at the proposal to repopulate Tasmania's wilderness, given that each cloned thylacine may have been worth millions of dollars. The need to protect them and the financial windfall to be had through displaying them suggested, instead, a form of captivity.

Captive thylacines did not breed, despite instances of males and females being kept together. And despite the optimism of some, such as Ronald C. Gunn, superintendent of convicts at Launceston, who in 1850 wrote of the pair destined for the Royal Zoological Society, 'I feel little doubt but that the Thylacines will do well and very probably breed . . . I can imagine nothing in the climate of London likely to injure them very materially'.[14] The chance of clones breeding naturally in captivity cannot be rated highly. That would suggest the need for artificial insemination (as is so often the case with the panda), removing further the reconstituted animal's naturalness.

Nick Mooney, former wildlife management officer with Tasmania's Parks and Wildlife Service and a long-time thylacine searcher and researcher, cautioned against the cloning project as a panacea: 'A snappy, technological quick-fix such as cloning extinct animals tempts many from the slog of fundamental environmental conservation.'[15]

There is also the ethical debate. Should scientists play God? Human beings caused the thylacine's extinction and shouldn't we have to live with that? At another level, there is the fundamental dilemma of 'copyright' over cloned specimens. Would they belong to the New South Wales Government? Archer's response as to whether steps had been taken to acquire copyright ownership was: 'Nothing yet. But somewhere down the line we've got to start looking at this sort of thing I suppose. We'll be taking sensible legal advice but my inclination has always been to keep the project and its potential benefits in the public sector.'[16]

The Australian Museum–Mike Archer project ended without a successful cloning, not least because the thylacine DNA was found to be contaminated with human DNA. Archer felt that the attempt had been worthwhile:

> It was ramping up the whole discussion about whether extinction really did have to be forever and whether we want any more species to go extinct. We needed to start down this path. Everybody recognises yesterday's science fiction can become tomorrow's science faction. We didn't know whether it would ever be able to be done. But all I was convinced of is, if we don't try it, we can be guaranteed it'll never happen.[17]

The lure of cloning the thylacine did not end there. In 2022 Colossal Laboratories & Biosciences, a breakthrough genetic engineering and de-extinction company, announced it had 'started the de-extinction of the thylacine, a beloved Australian marsupial that was eradicated by human hunting nearly a century ago . . . This is the second announced animal de-extinction project from Colossal, which uses breakthrough gene-editing technologies for a new wave of wildlife and ecosystem conservation.'[18]

Colossal was founded by US entrepreneur Ben Lamm and launched in 2021, its tagline being 'Extinction is a colossal problem facing the world'. Its first project is to genetically resurrect the woolly mammoth. Regarding the thylacine project, the lead researcher is Andrew Pask, University of Melbourne Professor in Epigenetics. Named 'moon-shot projects' for their high ambitions, the projects have raised significant funding. Pask's workplace is called the TIGGR Lab—Thylacine Integrated Genomic Restoration Research.

Pask apparently was inspired by the earlier Mike Archer project, calling him 'a pioneer and a big vision thinker who didn't have the genetic technology at the time to back up his big ideas'.[19] After learning about Archer's efforts, Pask and colleagues successfully brought a fragment of DNA from an extinct thylacine back to life in 2008, by editing a mouse embryo's genome and inserting the fragment into the embryo. The proposed full de-extinction is a four-step process, involving a tiny fat-tailed dunnart, the thylacine's closest living relative.

The process: (1) source stem cells from the dunnart; (2) edit the stem cells to match the genome of a thylacine; (3) remove the nucleus of a dunnart egg and replace it with the nucleus of the engineered thylacine stem cell; (4) implant the resulting embryo into a host. Timelines between step one and an actual birth are suggested by those involved to range from as little as five years to decades.

Colossal lists a five-point case for bringing back the thylacine: (1) a moral decency; (2) a boost to robust diversity; (3) ecosystem restoration; (4) rewilding Australia; (5) a scaleable, elegant endeavour with multidimensional ramifications.[20]

Prominent critics of the proposal include Kris Helgen, the Australian Museum's director and chief scientist, who formerly was curator of mammals at the Smithsonian Museum in Washington, DC. He believes the dunnart and thylacine are far too distantly related for the project to be anything other than fanciful, with a guarantee of raking in millions of dollars for a 'charismatic' animal de-extinction rather than aiming for a more realistic attempt with very closely related extinct animals.

Ethical and practical consequences of the de-extinction of *Thylacinus cyncocephalus* are complex, as noted earlier in the chapter. Will 'it' be copyrighted? Will 'they' be let free into wilderness in Tasmania and mainland Australia? Will 'they' breed in captivity and be displayed in zoos in Australia and elsewhere? And what should be done if the animal that is born is clearly not a thylacine?

Public perception will inevitably come into play again. Three years after the discontinuation of the Australian Museum cloning project, a thoroughly researched academic paper, 'Bring 'em back alive: Taming the Tasmanian tiger cloning project', was published in the Elsevier journal *Technology in Society*. Written by Amy Lynn Fletcher, Senior Lecturer in political science at the University of Canterbury in Christchurch, New Zealand, the paper assessed controversies surrounding the cloning project:

> Analysis of public documents (such as Museum annual reports) and media accounts (from Australian and international newspapers) demonstrates that the interest generated by the thylacine also destabilised the Museum's narrative of scientific progress and environmental restoration. The project kept morphing from science to spectacle and back again, becoming a theatre within which stakeholders could negotiate the ethical and political boundaries of research on conquering species extinction . . . While the Museum tried throughout the cloning project to position the work firmly within the boundaries of legitimate scientific research on paleogenomics, [the project] unleashed a fury of public debate that blurred the boundaries between science and politics, past and present (and future), and expertise and lay knowledge. Hovering over the debate throughout was the spectral image of the Tasmanian tiger, which exists both everywhere and nowhere in contemporary Tasmania.[21]

16

SIGHTINGS AND THE SCIENCE OF SURVIVAL

> I remembered how docile they seemed in captivity. It stood about twenty inches in height, with a heavy body and powerful looking jaws. The colouring was tawny, similar to the lion with dark brown stripes running from back to sides. It seems unlikely that there are any tigers left in Tasmania, but if any are found I trust they will be left in their natural habitat.
>
> MRS MYRA DRANSFIELD, ROKEBY

'It was embellished precisely because it could not yet be disproved.' Daniel J. Boorstin, US historian and Librarian of Congress during the 1970s, wrote those words in reference not to the Tasmanian tiger but to its entire habitat, *Terra Australis Incognita*, the southern land not yet discovered.[1] It was the fabled land that grew and shrank and grew again as maps changed across the centuries, until the day Abel Tasman, Ide Tjercxzoon and their shipmates dropped anchor.

Boorstin's statement is just as applicable to the thylacine, about which the big question, perhaps the only question left, is: could it still be out there?

If the thylacine still exists it will be miraculous, because there has been virtually no scientifically recognised physical evidence to support the case for its presence for the greater part of a century. Yet not only are there numerous instances of supposedly extinct species coming back—proof that science can be wrong—there have also been thousands of documented sightings of 'the tiger': in Tasmania, across mainland Australia and New Guinea. Either many people are wrong almost all of the time, or the thylacine is alive and well but in sufficiently low numbers, and sufficiently remote locations, to have avoided detection (despite many high-tech surveillance cameras in likely habitats).

To muddy the waters further, devil sightings continue to be reported in southern Australia and there is potential nineteenth-century evidence of thylacine activity in South Australia, suggested by Paddle, the Kimberley region, suggested by Guiler and Godard, and this incredible rather than credible 'sighting' in Queensland in 1890, as written up by E. Onstalet in *La Nature*:

> On two occasions the Zoological Society of London has received information that tends to make us believe in the existence in Australia of a carnivorous marsupial more or less analogous, if not identical, with the thylacine. In a letter addressed to Mr Sclater, Mr B. G. Sheridan, of Cardwell (Queensland), states, in fact, that his son, a boy of 13, who was accustomed to run the woods like an old hunter, was out one day accompanied by a small terrier, when he saw the latter obtain a scent and follow it up with eagerness. Curious to know what game he had to do with, the boy ran after his dog, and found himself face to face with an animal of the size of a dingo dog, with a round head like that of a cat, with a long tail, and with a body striped with yellow and black, and which

> was crouching in the high grass at about a mile from the coast. The dog and the savage beast soon grappled, and the boy, in order to aid his companion, tried to kill the enemy with a pistol shot, but having merely succeeded in wounding and rendering it more furious, he judged it prudent to beat a retreat. An animal of the same species was also perceived by a police officer of the same district, and traces of it have been observed on several occasions . . .[2]

Documented sightings excite the optimist in us, but need to stand up to rigorous testing to be considered viable. They also need to be weighed against current scientific opinion as to the likelihood of remnant populations still being out there. While the case against the thylacine's continued existence may appear overwhelming, there are nonetheless still some true believers and hunters.

In 1980, Tasmanian Parks & Wildlife Service zoologist Steven J. Smith published a detailed report on the status of the thylacine at that point. Included in his report was an analysis of every reported sighting from 1936 to the year of publication: 320 in all. That averages out at about one per month across a 24-year period, but the actual incidence was very different. Sightings increased decade by decade, gradually then dramatically, from just 21 in the 1940s to 125 in the 1970s.

Smith's report is important because there is no doubt that the very last of the species did *not* die in Beaumaris Zoo on 7 September 1936. Conservation biologists and other experts are in general agreement that the thylacine existed at least into the immediate post-World War II period—the time of fewest sightings, ironically when they were well under legislative protection. The presently unanswerable question is whether those remnant groups had reached the point of functional extinction,

APPENDIX B

REF. NO. | REF. NO.

NATIONAL PARKS AND WILDLIFE SERVICE

THYLACINE (TASMANIAN TIGER) SIGHTING REPORT FORM

OFFICE USE
G
F
P

Every observation, no matter how trivial it may seem, might prove to be important in the search for the thylacine. All information will be re-received in strict confidence. Thank you for your assistance.

NAME .... PHONE NO. ....

ADDRESS ....

(DIRECTIONS TO HOUSE) ....

OCCUPATION .... ACTIVITY AT TIME OF SIGHTING (e.g. shooting) ....

DATE REPORTED ...../..../19.... RECEIVED BY .... INTERVIEWED BY............................

SIGHTING DETAILS

LOCATION: Exact location; nearest reference point (map co-ordinates?).......

(Altitude ....m) ....

DATE ...../..../19... TIME .... HOW MANY MADE THE SIGHTINGS ....

TRANSPORT (car, foot, etc.) .... WHAT WAS SEEN (animal, tracks etc. ....

.... OBSERVED FOR (time) ....

IF ANIMAL SEEN, DESCRIBE ITS MOVEMENT (e.g. moving left, trotting etc.)..

....

....

WHERE SEEN (road or track/beach/paddocks, bush etc.).......................

....

SURROUNDING VEGETATION (Rainforest/gum trees/ low scrub/ buttongrass etc.)

....

.... DISTANCE (feet, metres, chains etc.)..............

LIGHT SOURCE (headlights, sun, moon etc.) ....WEATHER.................

DESCRIPTION OF ANIMAL (head, ears, body, tail, colour, markings, length nose to tail, height) ....

....

....

....

COMPARISON WITH OTHER BUSH ANIMALS (wild dog, wild cat, devil, wombat, tiger cat, native cat). Please comment briefly...............................

....

....

ANY OTHER COMMENTS ....

....

HAVE YOU SEEN A TIGER BEFORE ...... WHEN AND WHERE (alive, photo etc.)

....

NAMES AND ADDRESSES OF ANY OTHER IN THE PARTY (over page) ....

1 9 11 31 38 42 46 47 50 51 52 54 57 59 63 66 67 69 76 1 4

Sample page of The Tasmanian Tiger–1980: A report on an investigation of the current status of [the] thylacine, *Thylacinus cynocephalus,* funded by the World Wildlife Fund Australia, [by] Steven J. Smith, [published by] Tasmania: Department of Primary Industries, Parks, Water and Environment, May 1981, p. 121.

when a species is so rare as to no longer be influencing the ecology through selection, competition, predation, parasite spread and so on. Increased sightings suggested otherwise: but how believable were they? Each one went through many hoops before being given a final rating.

The parameters of the Smith survey seemed reasonably unambiguous:

> The sightings considered in this report are by people who claim that they saw, or may have seen a thylacine; or who saw an animal that they could not identify but whose description suggests that it may have been a thylacine. To be included in this sample the descriptions supplied by observers needed to be sufficient to identify a thylacine-like animal under the circumstances of the sighting . . . the criteria used to rate reports being the description of the animal, the observer's credibility and experience with native fauna, the circumstances of the sighting . . . and correlation with other sighting reports and previous thylacine distribution.[3]

However, thanks to human nature, the task would never be as simple as that:

> It is evident that the reporting of sightings is directly related to publicity given to searching for thylacines [and yet] many people are reluctant to make reports, for a variety of reasons, and it is known that a large number of sightings remain unrecorded. Some people fear that as a result of a report the location may be publicised and the animal endangered. Others fear that changes to the tenure and management of the locality [resulting from a good sighting] could threaten their livelihood or recreation; or fear personal ridicule.[4]

The Parks & Wildlife assessment form then in use was replaced by one based on an investigation into the Northern Rocky Mountain grey wolf. The new form exhaustively examined the phenomena of a sighting, to include: person by type (local, angler, shooter, hiker, etc.), observer's reliability (based on, for example, credibility within the community), location, length of time of sighting, number of people involved, mode of transport (including aircraft), nature of evidence (live, dead, tracks, scats, hair, call, fossil), description of animal including body colour, size, markings, distribution of stripes, head, tail, distance from animal (including more than 1 km), habitat (road, forest, beach, creek, etc.), light conditions (sun, moon, headlights, etc.), weather and visibility (including obscured vision), specific time of sighting, altitude, source of report (first-hand, second-hand, newspaper, museum, etc.), and correlation with other sightings since 1936.

Each element was rated. The rating system derived from the grey wolf survey form assigned a maximum of 10 points for the observer's reliability and credibility; a maximum of 25 points for the animal's description, increased points for a lengthy, close-up sighting, and so on, to enable a grand total to be reached. Smith's report found that 107 sightings were rated 'good', 101 'fair' and 112 'poor'. Twenty-plus years on, it's highly unlikely that a similar survey would rate sightings so generously. Not only has that much more time gone by without firm proof of the animal's continuing existence; the few experts in the state whose business it is to deal with sightings are necessarily far more rigorous in their assessments.

Even so, that hasn't stopped the sightings, with over a thousand in Tasmania to the present since 1936. In fact, as the body responsible for the thylacine, the Parks & Wildlife Service isn't universally admired for the stand it takes on the animal. Tiger hunters and enthusiasts beyond the public service tend to be referred to as 'true believers'; one of them, north-eastern Tasmanian identity Buck Emberg, went public about it:

> Now, perhaps, we can get the Parks & Wildlife people not to make tiger hunters and environmentalists like ourselves be seen as 'kooks' as we have in the past. I do not argue about whether the animal exists or not. We now have about 100 sightings from the past 25 years from dozens of people and no, we will not share it with the Parks & Wildlife people. They do not deserve our trust . . . yet. The animal will have to survive on its own. One day soon, we hope, verification will come. Until then we wish the perhaps five bands of tigers good luck and happy hiding.[5]

In his book *Thylacine: The Tragedy of the Tasmanian Tiger*, Eric Guiler wrote of an instance of a sighting 'widely believed to be true at the time [turning] out to be false'. This was an alleged aerial sighting on 6 January 1957 on a beach in the south-west, by helicopter pilot Captain Holyman and companion. A photograph was taken of the animal from the helicopter. Guiler examined the image and concluded that the animal was a dog, which seemed to be confirmed by the fact that a fisherman had lost his dog in the vicinity the previous month.[6]

A sighting two days earlier was presumably the same animal:

> The first live Tasmanian tiger to be seen for 27 years has been spotted from a helicopter. It was seen late on Wednesday [2 January] by Captain Jim Ferguson, co-pilot of the Australian National Airways helicopter, now based at Queenstown, Tasmania. The Tasmanian tiger is one of the most elusive creatures in the animal world, and was thought to have died out. Capt. Ferguson, with his flight engineer Mr Les Taylor, and Mr M. Solomon, a geologist with Mount Lyell Mining Company, were flying slowly home along the beach at Birthday

> Bay, 35 miles south-west of Queenstown, at a height of 30 ft., when they saw a tawny, tiger-striped 'dog' trotting along the beach. Realising it was no dog, Captain Ferguson whirled round and flew over the animal again—it WAS a Tasmanian tiger. Speaking from Queenstown last night, he said: 'We had a very clear view of the tiger for about two minutes. We have closely examined the natural history books here, and pictures, and we are convinced it was a tiger.' Argus wildlife expert Mr Crosbie Morrison said last night: 'This is wonderful news. The Tasmanian tiger was feared to be extinct. It is the largest-known flesh-eating marsupial and the fact it has been seen alive makes it the rarest mammal in the world.'[7]

As a result of this alleged sighting, Jim Ferguson is listed in the 1969 *Guinness Book of Records* as the last person known to have seen a live thylacine! Also strange is that Guiler made no mention of the Ferguson report in *The Argus* newspaper, one of the premier newspapers of the day in Australia.

Denuded of native habitat though it is, the north-east has long been a hotbed of sightings. It represented one-third of Smith's 'good' sightings and over one-third of the total sightings, with particular concentrations east and south of Golconda, where Emberg is based. Geographically, the north-east represents about 20 per cent of Tasmania's land surface. Whatever the cause, the conviction there is strong. Trevallyn resident Christine Lucas is a keen follower of the subject, arising out of personal experience. She is an example of an individual who has sighted but not reported. At the time of her encounter she had had but a 'passing interest' in the animal:

> I saw a tiger in Western Australia on New Year's Day 1991. I didn't report it for several reasons; they'd think I'd had a heavy night celebrating (I don't drink actually) and because at

This is a reproduction of the only photograph ever taken of a Tasmanian tiger. Note its rigid tail.

This photograph, which appeared in Pix *magazine, was taken of the animal sighted by Captain Holyman from his helicopter on 6 January 1957.* (Courtesy State Library of NSW)

> that point I had never heard of a tiger being sighted in the West . . . The one we saw was crossing a road as we came out of the Darling Range forests and into more open rural country heading from Perth north towards Northam. It seemed to be 'loping' across the road, was in no hurry. The only difference I saw was that the hind legs were not as tall as the one the artist drew [in *Australian Geographic* July–September 1986] . . . I can assure you I was not thinking Tassie Tiger at the time (on the way to visit family with a car load of our family too) and my 'imagination' has never seen one since! I was relating the experience to a friend here in Launceston and she said some years ago an old friend of theirs hit a Tassie Tiger while driving on the east coast of Tasmania. They were worried that they may have hurt it so reported it to the CSIRO in Hobart. They had just a little fur from their roo bar as evidence. Interestingly they were told not to speak to anyone about it. I am afraid I don't know the people or the year.[8]

In the north-west, true believer James Malley, who had formed the Thylacine Expedition Research Team with Jeremy Griffith and Bob Brown in 1972, remained an active and credible hunter and enthusiast. One sighting reported to him, of a thylacine chasing wombats, was considered to be the best for many years. As reported by *The Mercury* in the winter of 2002:

> Tiger hunter James Malley, who has spent almost half a century searching for a thylacine, said he had no reason to doubt the report. The man who said he saw the tiger is reluctant to go public because of the stigma associated with sightings . . . 'It was definitely a tiger, I get news of sightings like this extremely regularly and it all fits,' said Mr Malley, who immediately went to the area of the reported sighting. 'That's not the only one I have heard of in that area. Over the past two

> years I have probably had five and all fit with a seasonal pattern.' The man who reported the sighting said he had stopped his four-wheel-drive to engage its hubs and had turned the engine off. 'The wombats went past him for about fifteen metres, flat out into the bush,' said Mr Malley, who has scaled back the time he personally devotes to tiger searches. 'Then in front of him, the tiger was no further than five metres away. He was dumbfounded. The tiger stopped. He saw it for more than ten seconds and it just stopped and stared at him . . .' Mr Malley said he had been unable to find tiger footprints in the area [but] conditions were ideal for thylacines because there was plenty of game about but, like any animal facing extinction, the tigers were wary.[9]

Much further south, in the Styx and Florentine Valley area, Col Bailey—who passed away in 2022—lived and wrote about the thylacine, when not out searching. He believed that he first saw a thylacine in 1967 on the Coorong, the long coastal strip of south-east South Australia. His meeting in 1980 with the old Tasmanian fur-trapper, Reg Trigg, set Bailey on a course of writing short anecdotal accounts of thylacines, which were published in his local newspaper, the *Derwent Valley Gazette*. The column's popularity became such that it was syndicated worldwide and the tales were subsequently published as *Tiger Tales* (see Chapter 4). As a true believer, Bailey was able to write that his conviction was

> gained from over thirty years of personal experience in both researching and searching for this fascinating animal, yet it has often brought me into verbal conflict with less-opinionated citizens who loudly hammer the extinction drum. Be that as it may, it is my firm and unshakeable belief that the thylacine lives on.[10]

Well and good for faith. But, as history has shown elsewhere, faith and science don't always get along together. Dr Bob Brown once straddled that divide. The mystery of the thylacine had intrigued him since childhood and, for him, represented Tasmania. It was to be the lure delivering him to the island state. As a young man in 1972 he watched an ABC-TV *Four Corners* programme on the Lake Pedder controversy. It rekindled his interest and led to him taking what was to have been a short-term locum position in Launceston. That soon changed:

> When I came it wasn't a tourist's interest in the thylacine, it was actually a driven and active interest that I had . . . Yes, I ran into the Pedder people, but at the same time immediately started searching out information on the thylacine [and] quickly fell in with Jeremy and James . . . They were both glistening-eyed believers—they wouldn't hear that it was extinct. I was the sceptic. I was always the sceptic and said we had to look for evidence. There were all the sightings, but we had to look for evidence.[11]

Brown attributed his caution to six years of medicine 'and a few other myth-shattering experiences' he had behind him by then, resulting in a well-developed analytical mind. Over the eight-month period that the team was together they investigated some 250 sightings. For Brown, the complete lack of physical evidence—no scats, hairs, photographs, paw-prints—told only one story, but his colleagues remained otherwise convinced. Then, driving home to the Launceston suburb of Ravenswood one evening after his GP rounds, Brown saw a 'thylacine'. So astonished was he that he insisted Griffith return to the area with him immediately. Together, they found the animal: 'It was a greyhound with four stripes across its back'.[12]

There were just four sightings that the team couldn't explain. Brown's opinion is that they were probably wombats or dogs. His conviction that the thylacine is extinct extends to annoyance at deliberate, sustained misinformation surrounding it—not from fanciful sighters (Eric Guiler jested that sightings increase near pubs) but at the official level. In the 1960s, police, Guiler and others investigated a series of thylacine-like killings of sheep. They eventually trapped an animal that turned out to be an Alsatian dog. It was put down; the sheep killings stopped. But that was not made public until former police inspector Fleming recounted it to Brown.

Nor was the result of tests of hair samples of the famous 1961 thylacine kill by the two Sandy Cape fishermen who hid their slain thylacine under a metal sheet only to later discover that it had been stolen. At the time, Hobart CIB confirmed that the hairs belonged to a thylacine. A beekeeper, Reuben Charles, had also obtained some of the hairs, which he kept in a glass phial. Years after the event, Brown asked Charles for the phial and sent it to the Keith Turnbull Institute in Melbourne, which had acknowledged expertise in marsupial hair identification. The Institute reported that the hairs couldn't be positively identified, but were definitely not thylacine hairs.

In another instance, a Ben Lomond sighting yielded pawprints which turned out to belong to a wombat. Brown was of the opinion that:

> Science itself suppressed that information. They [scientists] led people out there to think, this [living thylacines] is happening. And it was wrong. It happened time and again . . . There were some decades in which it was almost disloyal, or unTasmanian, to reveal that these major sightings which had been echoed around the world were in fact false.[13]

The logical, scientific likelihood of the thylacine's being extinct is given further weight for Brown by looking not just at the present but the remote past; the thylacine had spent millions of years evolving and could not rapidly adapt to being targeted as vermin. He believes that, like any specialised predator, it has a range, and once moved off that range, that's the end. It is an argument that gives him no pleasure. But there may have been a twinkle in his eye when it was put to him that the animal's extinction remains unproven: 'That is very true. You can't also disprove that there isn't a herd of mammoths or Tasmanian emus out there'.[14]

Brown, Griffith and Malley produced a formal report of their 1972 research effort. Griffith wrote:

> In defence of those people who have claimed sightings, I should say that I cannot prove them wrong. There is no suggestion on my part that they have lied. In nearly all cases they honestly believe they have seen thylacine. I am not questioning their integrity, but rather their interpretation of what they have seen. Above all, they should not be condemned or ridiculed.[15]

Fifty-plus years on, a case in point is the 2016 Derwent Valley sighting, footage of which was released in 2017 (referred to in Chapter 3), in which a group called the Booth Richardson Tiger Team had set up fourteen cameras in remote bushland within 50 kilometres of Maydena. They moved the cameras regularly and in November 2016 captured footage of an animal 'barking and nosing their camera . . . "I don't think it's a thylacine, I know it's a thylacine," Adrian Richardson said.'[16]

The reason that they set up the cameras is because Richardson's colleague Greg Booth, a long-time thylacine researcher and hunter, claimed to have stumbled across a thylacine a few years earlier, on Good Friday in 2015, describing the animal:

> It had a really big head, a really long snout, it had a scar [on its head]. Its ears were pointed and it had white around the eyes with dark brown eyes set back ... It was sitting down and looking at me. I was about eight feet away from it ... I couldn't believe it. I couldn't sleep for days afterwards.[17]

While the 2017 footage attracted considerable attention, wildlife experts who analysed it discounted the notion that the animal was a thylacine. For Nick Mooney, the footage was 'better than other stuff I've seen [but] based on anatomy, movement, behaviour, size' he assessed it as likely to have been a spotted quoll. David Owen, who is not a wildlife biologist, was invited by Derwent Valley residents Dick Warner and Rob Clark to Joe Booth's Derwent Valley home to view the night footage on a very large TV screen in Booth's lounge room. Owen's conclusion, after many viewings, was that the vision was too distant and abrupt to confirm it as anything other than a quadruped with a tail, and that therefore the claim failed on that test alone.

As senior wildlife officer with the Parks & Wildlife Service, Nick Mooney was for many years was the state government's official point of contact for matters regarding the thylacine, including sightings. His involvement dated back to at least 1982, when he conducted a two-year search of an area of the Arthur River after his Parks & Wildlife colleague, Hans Naarding, experienced what is still considered to be the most authentic known sighting. At night with a torch, through rain, Naarding, a level-headed individual with considerable experience of native fauna, watched an adult thylacine for about three minutes. He described it as being in excellent condition with twelve black stripes on a sandy coat. Mooney constructed a series of sand traps for paw-prints, but came up with nothing.

Mooney rated a 1997 Surrey Hills sighting as virtually on a par with Naarding's, with three since then of sufficiently high quality to make him keep an open mind—just.[18]

Nick Mooney developed a rating method, derived in part from discussions with psychologists and police, which includes such details as chest flash, ear posture, leg length and the animal's motion, if the sighter had a good view. But Mooney's reconstructions of thylacine and some alleged fox sightings have led him to conclude that the error in sighting time, for instance, is usually 400 per cent, meaning that if someone said they saw the animal for ten seconds it was probably two or three seconds.

There are two other features relating to fauna in Tasmania which intrigue him. One is that he sees a potential correlation between the presence of endangered wedge-tailed eagles and the existence of thylacines:

> These species have very similar diets and both like secluded bits of old forest for breeding. Maybe thylacines take more adult wallabies, eagles more young, but the resource is spread similarly. Both are adapted for open forests although they can survive in other less suitable places if push comes to shove. I think the distribution of breeding wedgies would be an excellent indicator of thylacine potential.[19]

However, the flip side to this is something he readily acknowledges:

> There is so much food for [thylacines] now that they really should be pretty common if at all. The abundance of [meat-eating] devils is good oblique evidence of a lack of thylacine. One has to come up with a pretty bizarre set of circumstances to have it exist, considering the lack of specimens or credible

> evidence such as prints. If thylacines exist in remote or semi-remote areas there really should be dispersion into high food areas where they are more likely to be found. That's the problem—needing a bizarre (by definition unlikely) case to be made.[20]

The Tasmanian Museum and Art Gallery, formerly the Tasmanian Museum, has been closely associated with thylacine business since the mid-nineteenth century. Its former thylacine specialists, zoology curators David Pemberton and Kathryn Medlock, were closely involved in assessing sightings, as well as providing information and expert advice to individuals, film companies and scientific organisations worldwide, and the Tasmanian Museum and Art Gallery continues this work, as interest in the animal remains constant. Alleged sightings are on the decline and it is likely instead that cloning will now keep this most enigmatic of predators in the spotlight, as will strange twists such as the 2002 auction of the eight-skin thylacine rug. By pure coincidence the auction took place on Saturday, 7 September: Threatened Species Day in Australia, so named to commemorate the death of the last captive thylacine in Beaumaris Zoo.

The Tasmanian Museum and Art Gallery's holdings represent the world's most diverse thylacine collection, with skins, pouch young, skulls, mounts and footprints representing about 45 individuals. Medlock has identified material relating to 400-odd other specimens around the world, with relatively large collections at London's Royal College of Surgeons and Natural History Museum and also at Oxford. It's a regrettable indication of the unintended complicity of the then Tasmanian Museum, and other local bodies, in hastening the thylacine's demise through the export of live specimens when the animal was already scarce. A measure of the potential value of this scattered, mostly unseen, rare material was evident in the major exhibition that travelled nationwide in 2001. 'Tasmanian Tiger: The Mystery of the Thylacine', curated by Medlock, was seen by over half a million people. Media interest was constant, not least because Australian museums are not noted for taking their treasures on the road.

When this book was first published twenty years ago it quoted David Pemberton, who had been involved to a small degree with the Nick Mooney search around Togari junction and the Arthur River area, as having an 'open mind' on the question of the thylacine being extinct or extant. This is no longer the case for him because of the existence today of cheap, easy-to-procure and reliable surveillance cameras. Pemberton accepts that the plethora of cameras set across the Tasmanian landscape would have detected a thylacine population by now.

He recalls that during the 1980s when he was conducting extensive fieldwork on Tasmanian devils, he would set traps for them with a glimmer of hope of securing a thylacine. He also recollects the excitement of field biologists and rangers during the Mooney search, an excitement built on the highly plausible 1982 sighting by a very good field operator, Hans Naarding. Today, much has changed. Optimism in respect of its survival is all but gone.

But that thylacines were real and charismatic animals of the Tasmanian bush, having no need to be invented like Yeti or Bigfoot, suggests that sightings at least will keep them alive and well and safe where no one can harm them again.

NOTES

Chapter 1

1. *Journal of Discovery by me, Abel Jans Tasman, of a Voyage from Batavia for making discoveries of the Unknown South Land, 1642*, translated into English in J. Burney, *A Chronological History of the Voyages and Discoveries in the South Sea or Pacific Ocean, Part III, from the Year 1620, to the Year 1688*, London, Luke Hansard & Sons, 1813.
2. Saint-Hilaire had extensive knowledge of marsupials: in 1804 he took personal charge of the 40 000 preserved floral and faunal specimens brought back to France aboard the corvette *Geographe* from Baudin's epic voyage. He was on firm ground therefore when, in 1810, he showed that South America's marsupials were very different to the 'tyger'. He placed it in the entirely new sub-order Dasyuromorphia, representing carnivorous marsupials, from which derives the more common *dasyurus* (= shaggy tail; anglicised to dasyurid). Because this was a Greek-root term the Latin of the earlier classification had to change as well, and the animal became *Dasyurus cynocephalus*.

   In 1824 Coenraad Jacob Temminck refined the animal's classification further by differentiating it from the dasyurid family of quolls, dunnarts, antechinuses, planigales, ningauis and Tasmanian devils. It now belonged in its own unique Thylacinidae family (Greek *thylakos* = leather pouch). The third family in this group is the numbat. The fourth classified carnivorous marsupial, the blind marsupial mole, shares dasyurid, possum, kangaroo *and* placental mammal characteristics and, not surprisingly, is classified in its own sub-order. Although widespread in Australia's arid regions, and an excellent under-sand 'swimmer', not much is known about it.

An adjustment was made to Temminck's otherwise logical classification *Thylacinus cynocephalus*: his addition of the minor name *harrisii* was rejected. (The Tasmanian devil, the thylacine's closest relative, is *Sarcophilus harrisii.*)

3. Evans, George William, *Geographical, Historical and Topographical Description of Van Diemen's Land, with Important Hints to Emigrants*, London, John Souter, 1822, pp. 56–7.
4. Paddle, Robert, *The Last Tasmanian Tiger: The History and Extinction of the Thylacine*, Cambridge, Cambridge University Press, 2000, p. 7.
5. Troughton, Ellis, *Furred Animals of Australia*, Sydney, Angus & Robertson, 1941, 6th edn 1957, pp. 50 and 52.
6. Corbett, Jim, *Man Eaters of Kumaon Bombay*, Oxford University Press, 1944, p. xiv.
7. Morris, Desmond, *Animalwatching*, London, Jonathan Cape, 1990, p. 11.
8. University of Bristol research finding released 20 February 2023: 'Experts have discovered how zebra stripes work' www.bristol.ac.uk/news/2023/february/zebra-stripes.html, accessed 22 February 2023.
9. https://wolf.org/wolf-info/factsvsfiction/are-wolves-dangerous-to-humans/, report by John D. C. Linnell, Ekaterian Kovtun and Ive Rouart of the Norwegian Institute for Nature Research, accessed 22 February 2023.
10. Alderton, David, *Foxes, Wolves and Wild Dogs of the World*, London, Blandford, 1994, p. 141.
11. Penny, Malcolm, and Brett, Caroline, *Predators: Great Hunters of the Natural World*, London, Ebury, 1995, p. 98.
12. Guiler, Eric, *Thylacine: The Tragedy of the Tasmanian Tiger*, Melbourne, Oxford University Press, 1985, p. 39.
13. 'Towards the middle of the twentieth century, [the thylacine's] affinities again came into question. It had long been recognised as resembling a member of the wolf family but the similarities were clearly superficial and attributable to convergent evolution. However, it also bore a remarkable resemblance to the South American fossil

marsupial *Borhyaena* and strong arguments were advanced for regarding the thylacine as the sole Australian member of the Borhyaenidae. If this hypothesis had been correct, it would have had far-reaching implications regarding the origin of the Australian marsupials but recent intensive research has failed to provide evidence in favour of the South American relationship and has reinforced the location of the Thylacinidae in the Dasyuroidea. One of the most interesting researches involved the extraction of albumin from a piece of untanned skin of a thylacine and a serological comparison of this with albumins from diverse living marsupials. Results indicated only a very distant relationship of the thylacine to South American marsupials and a very close affinity with dasyurids. A further implication of these findings is that adaption to a running, predatory way of life has led *three* animal groups—Thylacinidae, Borhyaenidae and Canidae—to a similar functional solution.' (Rounsevell, D. E., in Ronald Strahan (ed.), *The Australian Museum Complete Book of Australian Mammals*, Sydney, Angus & Robertson, 1983, repr. with corrections 1988, p. 81.)

Chapter 2

1. https://pubs.geoscienceworld.org/gsa/geology/article-abstract/47/11/1039/573756/Nano-porous-pyrite-and-organic-matter-in-3-5, accessed 23 February 2023.
2. www.thylacoleo.com/whatis [accessed 10 July 2002]
3. Amphibious fishes required more than a benign atmosphere to survive on land. A number of critical evolutionary developments had already taken place in water, including the mysterious development of bone—enabling locomotion as well as protecting internal organs and the central nervous system—and jaws. Interestingly, the jaw was much later to become the principal evolutionary feature demarcating the first true mammals. In some creatures it began to become fused directly to the skull, thereby freeing up two little hinge-bones that became part of the middle ear, greatly enhancing hearing. The thylacine's jaw can open about 75°. Its famously wide gape is a threat-yawn; a display common to marsupial carnivores.

4. A much more widespread extinction 200 million years earlier marks the end of the Palaeozoic and the start of the Mesozoic. Here too, larger creatures, both terrestrial and aquatic, suffered most.
5. Archer, Michael, 'The dasyurid dentition and its relationship to that of didelphids, thylacinids, borhyaenids (Marsupicarnivora) and peramelids (Peramelina : Marsupialia)', *Australian Journal of Zoology*, *Supplementary Series*, January 1976.
6. Radiation refers to the evolutionary result of adaption and divergence. Thus, radiation by the mammal subclass *Marsupialia* has been tremendously successful and varied. To name some: dasyurids (quolls, dunnarts, planigales, devils, thylacines, numbats); perameloids (bandicoots, bilbies); vombatoids (koalas, wombats); phalangeroids (three possum families, cuscuses, gliders); macropods (potoroos, bettongs, kangaroos, wallabies, pademelons, quokkas). Kangaroos alone have radiated into about 60 species. At the other end of the scale the ancient monotremes display limited radiation, represented as they are by just two species of terrestrial echidna and the monospecific aquatic platypus.
7. Rolls, Eric, *Australia: A Biography*, St Lucia, University of Queensland Press, 2000, p. 108.
8. Convergent evolution occurs when unrelated animals adapt in similar ways to similar environments.

Chapter 3

1. Guiler, Eric and Godard, Philippe, *Tasmanian Tiger: A Lesson to be Learnt*, Perth, Abrolhos Publishing, 1998; Mooney, Nick, in *Thylacine: The History, Ecology and Loss of the Tasmanian Tiger*, edited by Branden Holmes and Gareth Linnard, p. vi.
2. Wilson, Edward O., *The Diversity of Life*, London Penguin, 1994, p. 34.
3. In the biological sense of the term, mammalian adaption takes place across thousands of generations. It manifests as speciation, 'the evolutionary process that throws off new species . . . As a species encounters fresh environments, brought about by factors such as climate change, it adapts, and so alters in different ways in

different parts of its range. Eventually a new form becomes differentiated enough to rank as a new species. The parent form, if unable to fit in with changed circumstances, disappears, while the genetic material persists, diversified and enriched'. (Norman Myers, *The Sinking Ark*, Oxford, Pergamon, 1979, p. 27.)

4. Ward, Peter, *End of Evolution: Dinosaurs, Mass Extinction and Biodiversity*, London, Weidenfeld & Nicolson, 1995, p. 252. Estimates of species extant at any time vary wildly, from 10 to 30 million. The only certainty is that insects comprise the vast majority, with millions of insect species becoming extinct before being known to science.
5. The dingo, *Canis familiaris dingo*, belongs to an equatorial group of primitive dogs and is not the same as the domestic dog, *Canis familiaris*, which for example breeds twice a year as opposed to the dingo's once. However they can, and do, interbreed.
6. Strahan (ed.), *Complete Australian Mammals*, p. 478.
7. https://wwfcee.org/news/wwfs-living-planet-report-reveals-a-devastating-69-drop-in-wildlife-populations-on-average-in-less-than-a-lifetime, accessed 27 March 2023.
8. *The Mercury*, Tuesday 28 March 2023, p. 2.
9. Holmes, Branden and Linnard, Gareth, eds, *Thylacine: The History, Ecology and Loss of the Tasmanian Tiger*, Melbourne, CSIRO Publishing, p. 169.

Chapter 4

1. Guiler, *Thylacine*, p. 67.
2. Paddle, *The Last Tasmanian Tiger*, p. 8.
3. *The Tasmanian*, 11 July 1885.
4. Holmes and Linnard, *Thylacine*, p. 21.
5. Holmes and Linnard, *Thylacine*, p. 8.
6. Rounsevell, cited in Strahan (ed.), *Complete Australian Mammals*, p. 82.
7. Paddle, *The Last Tasmanian Tiger*, p. 8.
8. Holmes and Linnard, *Thylacine*, p. 13.
9. Holmes and Linnard, *Thylacine*, p. 18.

10. Cited in Adam Smith, Patsy, *Tiger Country*, Sydney, Rigby Ltd, 1968, p. 55.
11. Strahan (ed.), *Complete Australian Mammals*, p. 175. Why do kangaroos and wallabies hop? 'There is no certain answer to the question but, inasmuch as this method of locomotion is found elsewhere among mammals only in small animals such as hopping-mice, it may be that the habit arose in small ancestors of the macropods and was retained because there was no great evolutionary pressure to revert to a more generally effective quadrupedal gait.' The writer does go on to say that, at speed, the bipedal gait is 'marginally more efficient in energy expenditure' than the other. It should be remembered that bipedalism also makes an animal taller—an argument used to explain why humans stood up—to keep watch for predators over long grass.
12. *Launceston Examiner*, 22 March 1899.
13. *The Mercury*, Thursday 9 September 1926, p. 16.
14. Mont Turner, cited in Adam Smith, Patsy, *Tiger Country*, p. 55.
15. Interview with the author, 14 August 2002.
16. Knight, Charles R and Knight, Annis Hardcastle, 'The Tasmanian Wolf', in *The Century Magazine*, Volume LXV1, No. 1, May 1903, New York, pp. 113–115.
17. *Advocate* (Burnie), Saturday 2 July 1927, p. 2.
18. Interview with the author, 14 August 2002. The Laird article appeared in *The Mercury*, 7 October 1968.
19. Holmes and Linnard, *Thylacine*, p. 10.
20. Guiler, *Thylacine*, p. 77.
21. van Lawick-Goodall, Hugo and Jane, *Innocent Killers*, London, Collins, 1970, pp. 61–2.
22. It seems that there has always been a broader argument in the Australian scientific community as to whether marsupial mammals are at a less developed evolutionary stage than the continent's placental mammals. If so, it illogically renders them biologically inferior: a decree apparently handed down to the colonials by nineteenth-century British scientists and not subsequently challenged to any vigorous degree. One implication of

this hierarchical ranking would have been a preparedness to accept their extermination, through being a badly adapted, primitive vermin species.

23. *Express and Telegraph* (Adelaide), Saturday 24 November 1900, p. 3. Article by John H. Carver.
24. Sharland, Michael, *Tasmanian Wild Life: A Popular Account of the Furred Land Mammals, Snakes and Introduced Mammals of Tasmania*, Melbourne, Melbourne University Press, 1962, p. 2.
25. Guiler, Eric, and Godard, Philippe, *Tasmanian Tiger*, pp. 183 and 206. Guiler has numerous photographs of animals believed to be thylacine kills. One is a small wallaby found in 1976: 'The throat, neck and upper chest have been ripped out but the remainder of the body is untouched. There has been no chewing or biting on any other part of the carcass'. Another is a sheep found in 1987: 'The carcass was undamaged except that the top of the skull was removed with the skin turned back as far as the neck. The brain had been eaten and there was no blood on or near the carcass'.
26. Guiler, Eric, *The Tasmanian Tiger in Pictures*, Hobart, St David's Park Publishing, 1991, p. 14.
27. Adam Smith, Patsy, op cit., p. 57.
28. *Empire* (Sydney), Friday 2 January 1863, p. 3.
29. *Telegraph* (Brisbane), Friday 24 March 1905, p. 5.
30. Gunn, Ronald, quoted in Paddle, *The Last Tasmanian Tiger*, p. 71, from *Papers and Proceedings of the Royal Society of Van Diemen's Land*, 2, pp. 156–7.
31. *Field*, Saturday 3 January 1885, p. 27.
32. ibid.
33. ibid.
34. Bailey, Col, *Tiger Tales: Stories of the Tasmanian Tiger*, Sydney, HarperCollins, 2001, p. 69.
35. *Circular Head Chronicle* (Stanley), Wednesday 6 October 1937, p. 3.

Chapter 5

1. Risdon Cove is of particular significance to the Tasmanian Aboriginal community. In May 1804 the members of a kangaroo hunting party of the Moomairemener tribe were attacked there by a small group of British soldiers, with an unknown number shot dead.
2. Giblin, R. W., *The Early History of Tasmania: the Geographical Era 1642–1804*, London, Methuen, 1928, p. 266. Whereas the streets of Hobart today are soberly named after its military and civic founders, Knopwood's Retreat is, appropriately, the city's most famous pub, in the heart of historic Salamanca Place.
3. Quoted in Evans, *Description of Van Diemen's Land*, p. 3.
4. Robson, Lloyd, *A History of Tasmania*, Melbourne, Oxford University Press, 1983, p. 46.
5. ibid. p. 70.
6. Boyce, James, *Van Diemen's Land*, Melbourne, Black Inc., 2008, p. 251.

Chapter 6

1. 'In the beginning was the Dreamtime and all things took their shape in the Dreaming. Trowenna [lutruwita], the heart-shaped island we call Tasmania, was very small, just a tiny sand-bank in the southern seas . . .'. So begins a traditional story of the creation of the island, one of a number told to the Quaker Joseph Cotton in the 1830s by Timler, an Aboriginal 'high priest'. Cotton's great-grandson adapted them from their original diary form. (*Touch the Morning: Tasmanian Native Legends*, by Jackson Cotton, Hobart, OBM Pty Ltd, 1979, p. 6.)
2. Clark, Julia, *The Aboriginal People of Tasmania*, Hobart, Tasmanian Museum and Art Gallery, 1986, p. 55.
3. George Augustus Robinson, a self-educated Londoner, migrated to Van Diemen's Land in 1824, working in Hobart Town as a builder, philanthropist and active Christian. In the wake of the devastation caused to the Aboriginal people through settler clashes and Governor Arthur's declaration of martial law, Robinson was appointed by Arthur to bring about conciliation with the remnant population.

He took to the task with extraordinary zeal, spending four years in the bush. By methods including inducement and capture for reward, he rounded up 135 Palawa and had them relocated to Flinders Island, there to 'save' and convert them. Unfortunately, disease, hunger and dislocation all but destroyed the pitiful group; only 47 survivors were moved back to Oyster Cover south of Hobart Town.

4. Cotton, *Touch the Morning*, p. 18.
5. Cotton, *Touch the Morning*, pp. 17–18. It is worth noting that, both in context and expression, the written, Anglicised Cotton stories undoubtedly differ from the original Aboriginal oral versions.
6. ibid., p. 23. Taraba's 'sneaky and wicked ways' are eventually punished when little bush elves make him ugly, whining and growling, then blow snowy white clematis leaves all over him to make him spotted, and rub wildcurrant juice on his ears to make them rosy red, as a sign that everyone speaks unpleasantly of him.
7. Unlike the bands, the tribes were not named and so have been given geographical titles. A band was made up of a number of hearth groups, each of these being a close-knit, extended family cooking around a single fire. A typical band might comprise 50 people, or seven or eight families.
8. Paddle, *The Last Tasmanian Tiger*, pp. 45–6.
9. Ryan, Lyndall, *The Aboriginal Tasmanians*, 2nd edn, Sydney, Allen & Unwin, 1996, pp. 34 and 38.
10. Barrett, Charles, *Isle of Mountains: Roaming Through Tasmania*, Melbourne, Cassell & Company, 1944, p. 28.
11. Guiler, *Thylacine*, p. 72.
12. Lourandos, Harry, *Continent of Hunter-Gatherers*, Cambridge University Press, 1997, p. 264.
13. ibid, p. 265.
14. Letter of Adam Amos, dated 20 April, 1826, quoted in Barrett, *Isle of Mountains*, p. 132.

Chapter 7

1. Jeffreys, C. H., *Geographical and Descriptive Delineation of the Island of Van Diemen's Land*, London, Richardson, 1820.

2. Robson, Lloyd, *A History of Tasmania Volume 1*, Melbourne, Oxford University Press, 1983, p. 86.
3. *Australasian* (Melbourne), Saturday 11 June 1892, p. 6.
4. Evans, *Description of Van Diemen's Land*, p. 28.
5. West, John, *The History of Tasmania Volume 1*, Launceston, Henry Dowling, 1852, p. 59 (Angus & Robertson reprint 1971).
6. Robson, *A History of Tasmania*, *Volume 1*, p. 95.
7. 'Woorady belonged to the Nuenone, a band of the South East tribe whose land was on Bruny Island. He was a very knowledgeable man and was renowned for his skills as a doctor, storyteller and boat builder. In 1829 his wife and one of his three sons died from illnesses introduced by Europeans. In 1830 he married Truganini and joined George Robinson. In 1835 he went with Truganini and his sons Myunge and Droyer-loinne to live at Wybalenna [on Flinders Island]. He was on the mainland of Tasmania from March 1836 to July 1837 searching for Aborigines who remained in the north-west. He returned to Wybalenna and in 1839 went to Victoria with George Robinson. Woorady died of old age in July 1842 while on board ship returning to Wybalenna from Victoria.' (*Living With the Land, Book Two: Resistance*, Hobart: Department of Education and the Arts, 1989, p. 38.) The name has various spellings.
8. 'Umarrah's tribal names were Kanneher Largenna and Molteher Largenna. He adopted the name Umarrah (an abbreviation of Hugh Murray) when he worked for Europeans who took over land near Campbell Town. He was a member of the Tyerer-note-panner, a band of the North Midlands Tribe and was an outstanding resistance leader. He was captured in November 1828 near the Eastern Marshes by Gilbert Robertson, the police chief from Richmond, whose roving party was guided by Kickerter-poller. Umarrah joined the group of Aboriginal people with George Robinson in January 1830 but left a few months later. In September he was either recaptured or surrendered and for a short while worked as a guide on the Black Line. He escaped and carried out raids on settlers along the Tamar River, before being recaptured in

August 1831. He rejoined George Robinson's search party but died of dysentery in Launceston on 24 March 1832.' (ibid., p. 38.)

9. Extracted from N. J. B. Plomley (ed.), *Friendly Mission: The Tasmanian Journals and Papers of George Augustus Robinson 1829–1834*, Hobart, Tasmanian Historical Research Association, 1966.
10. Meredith, Louisa Anne, *My Home in Tasmania, During a Residence of Nine Years*, London, John Murray, 1852 (quoted in Vivienne Rae-Ellis, *Louisa Anne Meredith: A Tigress in Exile*, Sandy Bay: Blubber Head Press, 1979, pp. 146–7).
11. Crawford, Patsy, *The Track Cutter: The Life of Thomas Bather Moore*, published by the author, 2020, pp. 34–35.

Chapter 8

1. Robson, Lloyd, *A Short History of Tasmania*, Melbourne, Oxford University Press, 1985, p. 33.
2. Report by Lt James Hobbs RN, 1824, quoted in Kerry Pink, *And Wealth For Toil: A History of North-West and Western Tasmania 1825–1900*, Burnie, Advocate Marketing Services, 1990, p. 23.
3. Letter of Edward Curr, 16 April 1830, quoted in Paddle, *The Last Tasmanian Tiger*, p. 110.
4. So-called necker snares 'consist of a noose of wire or hemp placed around a hole in a fence or a constriction in an animal trail in such a way as to strangle the animal when it became caught in the noose. The removal of one of the palings from the wire fences allowed game to pass through and at the same time provided an ideal snaring site'. (Guiler, *Thylacine*, p. 95.)
5. Margaret Howe, 'The Building and Rebuilding of Basil Steers #1 Hut: February Plains', Mountain Huts Preservation Society, p. 5.
6. Guiler, *Thylacine*, p. 105.
7. Robson, *A Short History . . .*, p. 30.
8. Gunn, Ronald, in West, *The History of Tasmania*, pp. 245–6.
9. Gould, John, in *The Best of Gould's Mammals*, Sydney, Macmillan, 1977, p. 48.
10. Robson, *A Short History . . .*, p. 51.

11. Fenton, James, *A History of Tasmania from its Discovery in 1642 to the Present Time*, Hobart, J. Walch & Sons, 1884, pp. 401–2.
12. Paddle, *The Last Tasmanian Tiger*, p. 143.
13. The wedge-tailed eagle, distributed throughout continental Australia, is one of the world's largest eagles. Tasmania's endemic subspecies is listed as 'endangered', having fewer than 200 breeding pairs. Wedge-tails feed on both live prey and carrion, with rabbits their preferred diet. Lambs make up only a small part of their total prey, but they have long been persecuted by sheep farmers and illegal poisoning and shooting continue in Tasmania.
14. *The Mercury*, 26 August 1884, p. 2.
15. *The Tasmanian Mail, Supplement,* 30 April 1887. Paddle has extensively researched this critical period; the Messrs French in particular he finds greatly exaggerated their thylacine kill claims.
16. Guiler, *Thylacine*, p. 22.
17. *The Mercury*, Monday 7 September 1874, p. 1.
18. *The Tasmanian News,* 6 September 1887.
19. *The Mercury*, 8 October 1886.

Chapter 9

1. *Davenport Leader* (Iowa, US), 3 March 1896, p. 5.
2. *Hobart Town Courier*, Saturday 28 February 1829, p. 2.
3. *The Examiner*, Saturday 19 April 1924, p. 7.
4. *Physical Description of New South Wales and Van Diemen's Land. Accompanied by Geological Map, Sections, and Diagrams, and Figures of the Organic Remains*, by P. E. de Strzelecki, London, Longman, Brown, Green, and Longmans, 1845, pp. 125–26.
5. *Illustrated Sydney News*, 16 April 1866, p. 14.
6. *Illustrated Adelaide Post*, Saturday 23 March 1867, p. 6.
7. *Cornwall Chronicle* (Launceston), Wednesday 14 December 1870, p. 2.
8. *The Sheffield & Rotherham Independent* (UK), Thursday 25 February 1875 p. 5. (Later appearing in *The Leicester Chronicle* and *Leicestershire Mercury* (UK), Saturday 19 August 1876, p. 11.)
9. *Australasian* (Melbourne), Saturday 24 May 1879, p. 24.

10. *Launceston Examiner*, Saturday 17 May 1884, p. 1.
11. William Bernhardt Tegetmeier writing in *Field*, Saturday 3 January 1885, p. 27.
12. *Launceston Examiner*, Thursday 31 January 1889, p. 3.
13. *Brooklyn Daily Eagle* (New York), 22 July 1894, p. 8.
14. *Australasian* (Melbourne), Saturday 31 July 1897, p. 29.
15. Knight, Charles R. and Knight, Annis Hardcastle, 'The Tasmanian Wolf', in *The Century Magazine*, volume LXV1, no. 1, May 1903, New York, pp. 113–115.
16. *Syracuse Sunday Herald* (Syracuse, US), 19 April 1903, p. 10.
17. *The Newsletter: An Australian paper for Australian People*, Sydney, Saturday 30 May 1903, p. 10.
18. *North Western Advocate and the Emu Bay Times* (Burnie), Wednesday 8 July 1908, p. 2.
19. *Meade County News* (Kansas, US), 23 September 1909, p. 2.
20. *North West Post* (Formby, Tasmania), Wednesday 24 February 1915, p. 4.
21. *Lebanon Daily News* (Lebanon, Pennsylvania, US), 30 July 1949, p. 6.
22. *Athens Sunday Messenger* (Athens, Ohio), 25 December 1960, p. 34.
23. *Uniontown Evening Standard* (Uniontown, Pennsylvania, US), 7 February 1964, p. 10.
24. *Blytheville Courier News*, 29 May 1980, p. 26.
25. Sighting by the well-respected wildlife biologist Hans Naarding, in Nick Mooney (Research Officer, Wildlife Division, National Parks and Wildlife, Tasmania) Search for Thylacines in Northwest Tasmania. Unpublished Research Project Report, 1983. Author's collection.

Chapter 10

1. *Express* (London), Friday 3 January 1851, p. 2.
2. *The Tasmanian*, 13 June 1885.
3. Knight, Charles R. and Knight, Annis Hardcastle, 'The Tasmanian Wolf', in *The Century Magazine*, Volume LXV1, No. 1, May 1903, New York, pp. 113–115.

4. *Hobart Mercury*, 5 December 1891.
5. Guiler, *Thylacine*, pp. 23 and 58.
6. *Weekly Courier*, 7 December 1911.
7. Robson, *A Short History . . .*, p. 70.
8. *Hobart Mercury*, 9 May 1893.
9. Atkinson, Henry Dresser, *The Woodpecker Papers: Talks with a Naturalist; Tales of Tasmania in the late 1800s*, [prepared by] Mary Hume, and Minnie Stein, Launceston, Bookprint, 2001, p. 44.
10. Smith, Geoffrey, *A Naturalist in Tasmania*, Oxford: Oxford University Press, 1909, pp. 95, 96–7.
11. Stephen R. Sleightholme and Cameron R. Campbell, 'A retrospective assessment of 20th century thylacine populations', *Australian Zoologist*, 38(1), 2016, pp. 102–129.
12. Guiler, *Thylacine*, pp. 24–5.
13. Norman Laird, article in *The Mercury*, 7 October 1968.
14. Paddle, *The Last Tasmanian Tiger*.
15. *Daily Telegraph* (Launceston), Monday 5 April 1909, p. 2.
16. *The Mercury*, Monday 31 May 1909, p. 6.
17. *Morning Astorian* (Oregon), Thursday 6 May 1909, p. 1. Victoria, B.C. 5 May.
18. *Daily Post* (Hobart), Wednesday 13 January 1909, p. 4.
19. *The Mercury*, 11 August 1908.
20. Cubit, Simon, *Snarers and Cattlemen of the Mersey High Country*, Launceston, Regal Press, 1987.
21. ibid., p. 34.
22. *Launceston Examiner*, Saturday 10 July 1897, p. 4.
23. Guiler, *Thylacine*, p. 115.
24. *Daily Telegraph* (Launceston), Monday 8 January 1906, p. 7.
25. *Australasian* (Melbourne), Saturday 31 July 1897, p. 29.
26. *The Mercury*, Saturday 30 December 1922, p. 11.
27. Article by Norman Laird in *The Mercury*, 7 October 1968.
28. Lord, Clive E., 'Notes on the mammals of Tasmania', in *Royal Society of Tasmania Papers and Proceedings*, 1917.
29. Lord, Clive E., 'Existing Tasmanian marsupials', in *Royal Society of Tasmania Papers and Proceedings*, 1926–28.

30. Le Souef, A. S. and Burrell, Harry, *The Wild Animals of Australasia*, London, Harrap, 1926, pp. 318–19
31. *Daily Post* (Hobart), Monday 12 October 1908, p. 3.
32. *The Mercury* Friday 23 April 1909, p. 4, quoting the *Daily Mail* (London), 18 March 1909.
33. *World* (Hobart), Wednesday 18 October 1922, p. 8.
34. *The Mercury*, Wednesday 1 August 1928, p. 7.
35. *The Examiner*, Wednesday 3 March 1937, p. 1.
36. *The Mercury*, Tuesday 9 March 1937, p. 2.
37. *The Examiner*, Wednesday 10 March 1937, p. 10.
38. *The Mercury*, Thursday 11 March 1937, p. 10.

Chapter 11

1. Robson, Lloyd, *A Short History . . .*, pp. 123, 126.
2. Paddle, *The Last Tasmanian Tiger*, p. 174.
3. ibid., p. 181.
4. Graves, Kathleen E., 'The rarest animal in the world', in *Walkabout* magazine, 1 April 1958, pp. 15–16.
5. Bell, E. A., ms paper NS 463/2, p. 7, Archives Office of Tasmania.
6. ibid., p. 11.
7. *The Mercury*, 22 January 1923.
8. *The Mercury, 70th Anniversary Special Issue*, 5 July 1924.
9. Bailey, Col, www.maydena.tco.asn.au/Colbail/tiger [accessed 11 January 2001.
10. Guiler, *Thylacine*, p. 55.
11. Paddle, *The Last Tasmanian Tiger*, p. 191.
12. Beresford, Quentin, and Bailey, Garry, *Search for the Tasmanian Tiger*, Hobart, Blubber Head Press, 1981.
13. www.tmag.tas.gov.au/whats_on/newsselect/2022articles/thylacine_mystery_solved_in_tmag_collections, accessed 18 March 2023.
14. Paddle, op. cit. p. 195.
15. Guiler, *Thylacine*, p. 29.
16. Conversation with the author, 2 October 2002.

Chapter 12

1. Letter of 30 May 1946, by N. J. B. Plomely, Director, Queen Victoria Museum, Launceston, to Mr Charles Rainbow, Nabowla, N. E. Coast, Tas.
2. Troughton, Ellis, *Furred Animals of Australia*, Sydney, Angus & Robertson, 6th edn., 1957, p. 52.
3. ibid., p.51.
4. Brogden, Stanley, *Tasmanian Journey*, Melbourne, Morris & Walker for Pioneer Tours, 1948 (repr. 1949), p. 79.
5. Robson, Lloyd, *A Short History* . . ., p. 172.
6. Guiler, Eric, and Godard, Philippe, *Tasmanian Tiger: A Lesson to be Learnt*, p. 190.
7. Guiler, *Thylacine*, p. 146.
8. Bell, E. A., *Thylacine Reports—Queen Victoria Museum—Launceston*, unpublished ms, Archives Office of Tasmania NS 896/1–39.
9. *The Mercury*, 15 April 1966.
10. *The Mercury*, 23 September 1966.
11. Burton, Bob, 'Wilderness and Unreasonable People', in Cassandra Pybus and Richard Flanagan (eds), *The Rest of the World is Watching: Tasmania and the Greens*, Sydney, Pan Macmillan, 1990, p. 80. (At the time of the article Bob Burton was a research officer for The Wilderness Society.)
12. *Giants of the Styx Forest State Forest in the Styx Valley and the Big Tree Reserve: A Guide for Visitors*, Forestry Tasmania pamphlet, March 2002.
13. Gee, H. M., 'The Evolving Consciousness', in Helen Gee and Janet Fenton (eds), *The South West Book: A Tasmanian Wilderness*, Sydney, Australian Conservation Foundation, 1978, p. 242.
14. Kiernan, Kevin, 'I Saw My Temple Ransacked', in Pybus and Flanagan (eds), *The Rest of the World*, p. 30. (Speliologist Kevin Kiernan helped found The Wilderness Society. He is also noted for rediscovering the Kutikina cave.)
15. Brown, Robert, 'Has the Thylacine Really Vanished?', in *Animals* 15, 1973, p. 419.
16. *Weekly Times* (Melbourne), Saturday 3 January 1903, p. 35.

Chapter 13

1. *Gastonia Gazette,* 20 August 1974, p. 15.
2. Beresford, Quentin, and Bailey, Garry, *Search for the Tasmanian Tiger*, Sandy Bay, Blubber Head Press, 1981.
3. *Tasmanian Mail*, 18 August 1981, p. 2.
4. Brown, Bob, 'Revelations of a Green Senator', in Helen Gee (ed.), *For the Forests: A History of the Tasmania Forest Campaigns*, Hobart, The Wilderness Society, 2001, p. 334.
5. *The Mercury*, 18 June 2002, p. 13.
6. Law, Geoff, interview with the author, 30 July 2002.
7. ibid.

Chapter 14

1. Statement supplied by Malcolm Wells, Deputy Chief Executive, Tourism Tasmania, 2 August 2002.
2. Gould, John, quoted in Dixon, Joan M., *The Best of Gould's Mammals*, Melbourne, Macmillan, 1984, p. 48.
3. *The Mercury*, 31 January 2002, p. 3.
4. Correspondence with the author, 5 August 2002.
5. Daniel Young, 23 March 2023, email to the author.
6. Smith, Nicholas, 'The return of the living dead: unsettlement and the Tasmanian tiger', *Journal of Australian Studies*, volume 36, no. 3, pp. 269–289, DOI: 10.1080/14443058.2012.703219
7. www.smh.com.au/national/much-valued-collaborator-on-art-projects-around-the-world-20211204-p59esv.html
8. Rouette, Georgia, '*Thylacinus Cynocephalus*: Lessons from Art', in *The World of Antiques and Art*, July–December 2002, p. 143.
9. Cochrane, Grace, catalogue notes for '*Thylacine: Pest to Pedestal. Recent Work* by Michael McWilliams', Lauraine Diggins Fine Art, North Caulfield, 2001. The exhibition of 55 works ran from 20 November to 15 December 2001.
10. ibid.
11. Hay, Peter, 'Lagunta', in *Story teller*, Hobart [nd].
12. Correspondence with the author, 13 August 2002.
13. Interview with the author, 31 July 2002. *The Tragedy and Myth of the Tasmanian Tiger* was produced by Roar Film Pty Ltd in association

with Screen Tasmania, with the assistance of the Tasmanian Museum and Art Gallery and G3 Multimedia, Hobart, 2001. Enquiries: info@roarfilm.com.au.

14. Correspondence with the author, 15 August 2002.
15. https://catalogue.nla.gov.au/Record/82056, accessed 25 February 2023.
16. Hadfield, Warwick, 'Burning Bright', in *Island* 88, reviewing *Tyger Tyger*.
17. Correspondence with the author, 20 August 2002.
18. Peek, Andrew, 'The Last Thylacine', in *Australian Book Review* No. 211, p. 30.

Chapter 15

1. Mike Archer, correspondence with the author, 4 August 2002.
2. www.aljazeera.com/features/2023/1/29/back-from-extinction-resurrecting-the-tasmanian-tiger, accessed 25 February 2023.
3. Maurice Paleau, an executive of the Discovery Network cable television company, quoted on the Australian Museum's website www.austmus.gov.au/thylacine [accessed 1 August 2002]. Discovery made a documentary, *End of Extinction: Cloning the Tasmanian Tiger*, which aired in 155 countries.
4. Mike Archer, correspondence with the author, 4 August 2002.
5. *The Thylacine Museum*, www.naturalworlds.org/thylacine/more topics/cloning [accessed 1 August 2002].
6. Mike Archer, correspondence with the author, 4 August 2002.
7. *Catalyst*, ABC Television, Thursday 21 February 2002.
8. ibid.
9. ibid.
10. Mike Archer, correspondence with the author, 4 August 2002.
11. ibid.
12. ibid.
13. Abraham Passmore, letter to *The Mercury*, 31 May 2002.
14. Gunn, Ronald C., letter dated 12 November 1850, to D. W. Mitchell, Secretary of the Royal Zoological Society. Quoted in *The Best of Gould's Mammals*, p. 48.

15. Letter to *The Mercury*, 4 June 2002.
16. Mike Archer, correspondence with the author, 4 August 2002.
17. www.aljazeera.com/features/2023/1/29/back-from-extinction-resurrecting-the-tasmanian-tiger, accessed 25 February 2023.
18. www.businesswire.com/news/home/20220816005043/en/Colossal-to-De-Extinct-the-Thylacine-also-known-as-the-Tasmanian-Tiger-an-Iconic-Australian-Marsupial-That-Has-Been-Extinct-Since-1936, accessed 26 February 2023.
19. www.aljazeera.com/features/2023/1/29/back-from-extinction-resurrecting-the-tasmanian-tiger, accessed 25 February 2023.
20. https://colossal.com/the-thylacine-an-exemplary-candidate-for-de-extinction/, accessed 26 February 2023.
21. Fletcher, Amy Lynn, 'Bring 'Em back alive: Taming the Tasmanian tiger cloning project', in *Technology in Society*, volume 30, issue 2, April 2008, Amsterdam, Elsevier, pp. 194–201.

Chapter 16

1. Boorstin, Daniel J., *The Discoverers*, Harmondsworth, Penguin, 1986, p. 278.
2. *Orange Leader* (New South Wales), 26 April 1890, p. 8.
3. Smith, Steven J., *The Tasmanian Tiger—1980: A Report on an Investigation of the Current Status of the Thylacine* Thylacinus cynocephalus, *Funded by the World Wildlife Fund*, Hobart: National Parks & Wildlife Service, 1980, p. 45.
4. ibid, pp. 45, 46.
5. Letter to *The Mercury*, 31 May 2002.
6. Guiler, *Thylacine*, p. 135.
7. *Argus* (Melbourne), Friday 4 January 1957, p. 1.
8. Letter to the author, April 2002.
9. Report by Luke Sayer in *The Mercury*, 2 June 2002, p. 9.
10. Bailey, Col, *Tiger Tales: Stories of the Tasmanian Tiger*, Sydney, HarperCollins, 2001, p. viii.
11. Interview with the author, 16 August 2002.
12. ibid.
13. ibid.

14. ibid.
15. *The Report of the Search for the Thylacine that was conducted by Jeremy Griffith, James Malley and Robert Brown*, [S.l, s.n.], December 1972, p. 19.
16. www.abc.net.au/news/2017-09-06/tasmanian-tiger-sighting-claimed-by-trio/8877598, accessed 16 April 2023.
17. ibid.
18. From *The Mercury*, 2 June 2002, 'Recent Tassie tiger sightings', p. 17: '*August 1997*—A motorist sees "thylacine" near road at Surrey Hills, behind Burnie. The driver braked suddenly to see it about three metres away. "Hair was short and dark brown with clear chocolate-coloured stripes." Watched the animal for about a minute.' '*December 1998*—A motorist sees "thylacine" near Pyengana. The driver described the animal as "dark tan with obvious stripes and a straight tail".' '*January 1999*—Two Victorian trout fishermen driving near Great Lake see "thylacine". They said it was "Golden-coloured", "knee-high" and "very long" with "clear rump stripes". They returned the next day to collect fur off fence-wire, droppings and footprints from the scene. Gave the evidence to authorities.'
19. Correspondence with the author, 20 August 2002.
20. ibid.

Select Bibliography

Adam Smith, Patsy, *Tiger Country*, Sydney, Rigby Ltd, 1968.

Alderton, David, *Foxes, Wolves and Wild Dogs of the World*, London, Blandford, 1994.

Atkinson, Henry Dresser, *The Woodpecker Papers: Talks with a Naturalist; Tales of Tasmania in the late 1800s*, [prepared by] Mary Hume and Minnie Stein, Launceston, Bookprint, 2001.

Bailey, Col, *Tiger Tales: Stories of the Tasmanian Tiger*, Sydney, HarperCollins, 2001 (reproduced by permission of the author).

Beresford, Quentin, and Bailey, Garry, *Search for the Tasmanian Tiger*, Hobart, Blubber Head Press, 1981.

Boorstin, Daniel J., *The Discoverers,* Harmondsworth, Penguin, 1986.

Cotton, Jackson, *Touch the Morning: Tasmanian Native Legends*, Hobart, OBM Pty Ltd, 1979.

Cubit, Simon, *Snarers and Cattlemen of the Mersey High Country*, Launceston, Regal Press, 1987.

Evans, George William, *A Geographical, Historical and Topographical Description of Van Diemen's Land*, London, Heinemann, 1822 (facsimile reprint 1967).

Fenton, James, *A History of Tasmania from its Discovery in 1642 to the Present Time*, Hobart, J. Walch & Sons, 1884.

Gee, Helen and Fenton, Janet (eds), *The South West Book: A Tasmanian Wilderness*, Sydney, Australian Conservation Foundation, 1978, p. 242.

Gee, Helen (ed.), *For the Forests: A History of the Tasmanian Forest Campaigns*, Hobart, The Wilderness Society, 2001, p. 54.

Giblin, R. W., *The Early History of Tasmania: The Geographical Era 1642–1804*, London, Methuen, 1928.

Guiler, Eric, *The Tasmanian Tiger in Pictures*, Hobart, St David's Park Publishing, 1991.

Guiler, Eric, *Thylacine: The Tragedy of the Tasmanian Tiger*, Melbourne, Oxford University Press, 1985 (reproduced by permission of Oxford University Press Australia).

Guiler, Eric and Godard, Philippe, *Tasmanian Tiger: A Lesson to be Learnt*, Perth, Abrolhos Publishing, 1998.

Hogan, T. M. and Gye, Hal, *The Tight Little Island*, Hobart, J. Walch & Sons [1914?].

Jeffreys, C. H., *Geographical and Descriptive Delineation of the Island of Van Diemen's Land*, London, Richardson, 1820.

Le Souef, A. S. and Burrell, Harry, *The Wild Animals of Australasia*, London, Harrap, 1926.

Living With the Land, Book Two: Resistance, Hobart, Department of Education and the Arts, 1989, co-ordinated by Heather Felton.

Lourandos, Harry, *Continent of Hunter-Gatherers*, Cambridge University Press, 1997.

Morris, Desmond, *Animalwatching*, London, Jonathan Cape, 1990.

Myers, Norman, *The Sinking Arc*, Pergamon, 1979.

Paddle, Robert, *The Last Tasmanian Tiger: The History and Extinction of the Thylacine*, Cambridge, Cambridge University Press, 2000 (reproduced by permission of the author and Cambridge University Press).

Penny, Malcolm and Brett, Caroline, *Predators: Great Hunters of the Natural World*, London, Ebury, 1995.

Pink, Kerry, *And Wealth For Toil: A History of North-West and Western Tasmania 1825–1900*, Burnie, Advocate Marketing Services, 1990.

Pybus, Cassandra and Flanagan, Richard, *The Rest of the World is Watching: Tasmania and the Greens*, Sydney, Pan Macmillan, 1990.

Rae-Ellis, Vivienne, *Louisa Anne Meredith: A Tigress in Exile*, Sandy Bay, Blubber Head Press, 1979.

Robinson, George Augustus, *Friendly Mission: The Tasmanian Journals and Papers of George Augustus Robinson, 1829-1834*, edited by N. J. B. Plomley, Hobart, Tasmanian Historical Research Association, 1966 (reproduced by permission of THRA and the Queen Victoria Museum and Art Gallery).

Robson, Lloyd, *A History of Tasmania Volume I*, Melbourne, Oxford University Press, 1983 (reproduced by permission of Oxford University Press Australia).

Robson, Lloyd, *A Short History of Tasmania*, Melbourne, Oxford University Press, 1985 (reproduced by permission of Oxford University Press Australia).

Ryan, Lyndall, *The Aboriginal Tasmanians*, 2nd edn, Sydney, Allen & Unwin, 1996.

Smith, Geoffrey, *A Naturalist in Tasmania,* Oxford, Oxford University Press, 1909.

Smith, Steven J., *The Tasmanian Tiger—1980: A report on an investigation of the current status of the thylacine* Thylacinus cynocephalus, *funded by the World Wildlife Fund,* Hobart, National Parks & Wildlife Service, 1981.

Strahan, Ronald (ed.), *The Australian Museum Complete Book of Australian Mammals*, Sydney, Angus & Robertson, 1983, reprinted with corrections 1988 (reproduced by permission of the Australian Museum).

The Best of Gould's Mammals, with modern commentaries by Joan M. Dixon, Sydney, Macmillan, 1977.

Troughton, Ellis, *Furred Animals of Australia*, Sydney, Angus & Robertson, 1941, 6th edn 1957 (reproduced by permission of HarperCollins).

van Lawick-Goodall, Hugo and Jane, *Innocent Killers*, London, Collins, 1970.

Ward, Peter, *End of Evolution: Dinosaurs, Mass Extinction and Biodiversity*, London, Weidenfeld & Nicolson, 1995.

West, John, *The History of Tasmania Volume I*, Launceston, Henry Dowling, 1852 (Angus & Robertson reprint 1971).

Wilson, Edward O., *The Diversity of Life,* Penguin, 1992.

INDEX

Images are in italics.